— PRAISE FOR

"*Eager* is a revelation! If we only let them live, bea[v]
ecological problems. Ben Goldfarb's wonderful [...]
these intelligent, inventive, resilient rodents than (if you have any sense) you are already —
and might just tail-slap a politician or two into realizing how much we need them to restore
our critical wetlands." — **Sy Montgomery**, author of *The Soul of an Octopus*
and coauthor of *Tamed and Untamed*

"Beavers are easy to caricature, and they're a bit comical. But they've got their serious
side, too. European settlers who cut, plowed, and shot their way west also trapped the
country nearly clean of mammals. Almost killing off beavers — the continent's major water
engineers and dam builders — caused widespread problems for wildlife and people. Now,
though, beavers are on the rebound, and the how and who of that story, as told in *Eager*, will
give you a new and completely different concept of the continent."

— **Carl Safina**, author of *The View From Lazy Point* and *Beyond Words*

"This witty, engrossing book will be a classic from the day it is published. No one who loves
the landscape of America will ever look at it quite the same way after understanding just how
profoundly it has been shaped by the beaver. And even the most pessimistic among us will feel
strong hope at the prospect that so much damage can be so easily repaired if we learn to live with
this most remarkable of creatures." — **Bill McKibben**, author of *The End of Nature*

"*Eager* is the stunning story of beavers — so integral to early human landscapes of North
America — and their function in support of people and later the American economy. Literally
nature's "Corps of Engineers," beavers today play vital roles in restoring watersheds, landscapes, and flood control throughout the continent. To view them just as a cute animal with
a flat tail is to trivialize a central player in both history and modern day landscape ecology."

— **Thomas E. Lovejoy**, University Professor of Environmental Science
and Policy, George Mason University

"*Eager* brilliantly presents the role of the American beaver in shaping the landscape of our continent and preserving its ecological integrity and diversity — and does so in clear, readable prose.
My Native ancestors — before the cultural disruptions of the fur trade — saw the beaver people as
a nation worthy of the greatest respect. I believe that any thoughtful person who reads this book
will come away with a much deeper appreciation of this sacred being's place in the America of the
past and, we hope, the future." — **Joseph Bruchac**, coauthor of *Keepers of the Earth*

"Long trapped for their fur and maligned as pests, beavers are finally recognized for their
role in keeping water in the landscape. Goldfarb's spirited, well-researched account tells
the story of humanity's relationship with beavers and highlights innovative efforts to ally
with them to restore rivers and wetlands and boost ecological resilience. Our winsome,
paddle-tailed friends could have no better champion."

— **Judith D. Schwartz**, author of *Cows Save the Planet* and *Water in Plain Sight*

"There are a number of books that focus on a single species, but the amazing story of the beaver, as told by Ben Goldfarb, is in a class all its own. Dear reader, prepare yourself to be awed by a rodent!"   — **Tom Wessels**, author of *Reading the Forested Landscape* and *Granite, Fire, and Fog*

"One of the best things that can be said about a book is that it is both necessary and good. Not many are, but this one is."   — **Richard Manning**, coauthor of *Go Wild*

"With the perfect blend of science and storytelling, Ben Goldfarb takes us on a remarkable journey to discover the myriad ways beavers have shaped our landscapes and history — and, if we are willing, could help us fix our broken water cycle. An absorbing and eye-opening book that comes at a crucial time."   — **Sandra Postel**, author of *Replenish*

"In *Eager* Ben Goldfarb demonstrates that beavers are more than just a fascinating and mysterious rodent — they're also an 'animal that doubles as an ecosystem.' Optimistic and exciting, the book suggests a future where rather than destroying nature, or trying to dominate it with heavy-handed management, we collaborate with species like beavers to create a wilder, more diverse, and surprising world. *Eager* will make a Beaver Believer out of you!"

— **Emma Marris**, fellow at the UCLA Institute of the Environment and Sustainability; author of *The Rambunctious Garden*

"Beavers do matter. Contrary to the popular image of beavers as trouble-making 'varmints' on the land, these hardworking animals play many critical roles in nature, including rewetting creeks in dry country. That might seem counterintuitive — beavers are famous dam builders after all — but as Ben Goldfarb explains in his riveting new book, the engineering prowess of these mighty rodents is essential to healthy riparian areas. And they do their work for free!"

— **Courtney White**, author of *Grass, Soil, Hope* and *Two Percent Solutions for the Planet*

"An important and engaging book about the nature of beavers, the forces of nature, and the hubris of humans. While I've read many books about how *Homo sapiens* extirpated species around the globe historically, and how we've wiped out birds such as turkeys and beasts such as bison and elk in the recent past, I had not read a book about beavers. This book is an eye-opening contribution with great examples of the power of beavers to restore ecosystems."

— **Fred Provenza**, author of *Nourishment*

"In this beautifully written tribute to beavers, Ben Goldfarb paints a vivid and captivating portrait of two of nature's most fascinating species, *Castor canadensis* and *Castor fiber*. Seamlessly combining history, ecology, biology, politics, and compelling stories of those battling over the proper role of beavers in today's anthropocentric world, *Eager* resoundingly proves that these magnificent rodents do indeed matter a great deal. In so doing, this gem of a book offers hope not only for the beavers' future, but also our own."

— **Eric Jay Dolin**, author of *Fur, Fortune, and Empire* and *Black Flags, Blue Waters*

Eager

Eager

THE SURPRISING, SECRET LIFE OF

BEAVERS

AND WHY THEY

MATTER

Ben Goldfarb

FOREWORD BY DAN FLORES

CHELSEA GREEN PUBLISHING
WHITE RIVER JUNCTION, VERMONT
LONDON, UK

Copyright © 2018 by Ben Goldfarb.
All rights reserved.

Unless otherwise noted, all photographs by Ben Goldfarb.
Illustrations on pages 29 and 71 copyright © 2018 by Sarah Gilman.

No part of this book may be transmitted or reproduced in any form by any means without permission
in writing from the publisher.

Grateful acknowledgement is given to *High Country News*, which originally published "The Beaver
Whisperer" and portions of "Wolftopia," and to *bioGraphic*, which published portions of "Dislodged."

Project Manager: Sarah Kovach
Developmental Editor: Michael Metivier
Copy Editor: Laura Jorstad
Proofreader: Rachel Shields Ebersole
Indexer: Ruth Satterlee
Designer: Melissa Jacobson

Printed in Canada.
First printing June 2018.
11 10 9 8 7   23 24 25 26 27

Our Commitment to Green Publishing

Chelsea Green sees publishing as a tool for cultural change and ecological stewardship. We strive to align
our book manufacturing practices with our editorial mission and to reduce the impact of our business
enterprise in the environment. We print our books using vegetable-based inks whenever possible. This
book may cost slightly more because it was printed on paper that contains recycled fiber, and we hope
you'll agree that it's worth it. *Eager* was printed on paper supplied by Marquis that is made of recycled
materials and other controlled sources.

Library of Congress Cataloging-in-Publication Data
Names: Goldfarb, Ben (Environmental journalist), author.
Title: Eager : the surprising, secret life of beavers and why they matter / Ben Goldfarb ;
   foreword by Dan Flores.
Description: White River Junction, Vermont : Chelsea Green Publishing, 2018.
   | Includes bibliographical references and index.
Identifiers: LCCN 2018004621 | ISBN 9781603587396 (hc) | ISBN 9781603589086 (pbk.)
   | ISBN 9781603587402 (ebook) | ISBN 9781603588386 (audiobook)
Subjects: LCSH: Beavers — Ecology — North America. | Wildlife conservation — North America.
   | Beavers — Habitations — North America — History. | Colonization (Ecology) — North America.
Classification: LCC QL737.R632 G64 2018 | DDC 333.95/937 — dc23
LC record available at https://lccn.loc.gov/2018004621

Chelsea Green Publishing
White River Junction, Vermont, USA
London, UK
www.chelseagreen.com

To LS and DG,
who provided the best example
of mated pair behavior a kit could ask for.

*Water is important to people who do not have it,
and the same is true of control.*

— JOAN DIDION, *THE WHITE ALBUM*, 1979

Contents

An America Without Beavers

I f you're like me, at some point as you read it, the book in your hands is going to send you outside to gaze across the landscape toward the nearest river valley. In my case the valley would be that of the Rio Galisteo, named after one of the long-ago Puebloan Indian towns that once dotted this broad drainage in the southwestern piñon-juniper country just south of Santa Fe, New Mexico. And the reason a new book about beavers has had me shading my eyes and looking off at the cottonwood corridor a mile away is because I suddenly have a potential new explanation for an old historical mystery. That may not be the motive of most readers of *Eager*, but I stand by the assertion that this book is going to make you look out on the world and see our wildlife story with new eyes.

If there is a popular saint of ecology in this country, it is Aldo Leopold. As the faithful know, after attending Yale Forestry School, Leopold spent his first years in the field in New Mexico. One of the places here where he obviously invested time was the valley where I live, and in a couple of his essays — "The Virgin Southwest" and "Pioneers and Gullies" — he employed the Rio Galisteo as an example, and not a pretty one. He used the Galisteo as Exhibit A of a western stream destroyed by gullying and told a before-and-after story about my local river to make his point. In 1849, he wrote, a drunken immigrant had been able to cross the Rio Galisteo successfully on a twenty-foot board plank. Yet by the early twentieth century, the Galisteo's waters had sliced its streambed into arroyos so resembling a rat maze that the drunk wouldn't stand a snowball's chance of making the far bank. A twentieth-century plank across the Rio Galisteo would have had to span 250 feet.

As far as Leopold could see, the culprits for this crime were still on the scene: The cow, the sheep, and the horse had destroyed the Rio Galisteo. But Ben Goldfarb's *Eager* has me rethinking Leopold's explanation. The great naturalist arrived too late in New Mexico to see what else had happened to the Rio Galisteo in an earlier age. Heading on the flanks of Thompson Peak in the Sangre de Cristos, the mountain range looming above Santa Fe and stretching into Colorado, the Galisteo was one of scores of Southern Rocky Mountain rivers that trappers picked clean of beavers in the 1820s. By the early 1830s desperate mountain men even fanned out across the nearby High Plains in a mostly fruitless search to excavate every last dam, beaver pond, and kit in the Southwest. Domestic animals, as Goldfarb instructs us, then hit the vegetation of riparian zones so hard that beavers simply couldn't reestablish.

Ignore for a moment the selfish-gene cause, the exploitation of a wild animal for mere fashion. These acts of beaver removal, as Goldfarb describes so vividly in the pages to follow, abruptly terminated centuries stretching deeply into the past when beavers and their works fashioned the continent's watercourses into ribbons of inundation and trickling water storage. With beavers gone, that wet world — the kind of wetness many places will long for in the coming climate — has yielded to flashier runoffs that have cut gullies and arroyos and helped produce a drier North America.

I first encountered Ben Goldfarb's work reading his superb, eye-catching articles in the famous western outdoor magazine, *High Country News*. Friends and I thought the quality of both his research and his writing heralded the arrival of a new voice on environmental issues, and this book has only cemented my belief. *Eager* tells its story in the vein of modern environmental journalist-writers Elizabeth Kolbert, Ian Frazier, and David Quammen. Which is to say that the reader gets to travel widely, both geographically and temporally; meet scores of fascinating people; and acquire sophisticated natural history and ecological knowledge by accompanying the author's inquisitive, generous questing into the world. Goldfarb's pursuit of the beaver's reign, its disastrous decline, and now the beginnings of its return will take readers of this book across pretty much all of the continental United States, from New England and the Hudson Valley to the Pacific Northwest, from the Northern Rocky Mountains to Utah and the Southwest. He even tracks the beaver's return in the Old World. Across America the author is dogged in searching out both landscapes and individuals important to the beaver's story, and especially the Beaver Believers who are shaping the animal's (still) unfolding biography. *Eager* is something close to a beaver walkabout on which a reader gets to tag along.

This is twenty-first-century nature writing at its best, and I for one am going to be very interested to see how the reading public reacts to Goldfarb's clearly delineated hope that we allow, indeed that we assist, beavers' return. I say that because the author has convinced me that beavers create a continent quite different from the one we might hold in our romantic imaginations about "virgin" (or Indian) America. Beaver America was — is — a swampier, boggier, muddier landscape than we might think. The sparkling, free-running mountain streams of modern fly-fisher or river-runner fantasies, much like the predator-free paradise of twentieth-century sport hunting, necessarily give way to a very different world when beavers and wolves are back.

But then, that kind of "entire heaven and entire earth," as Thoreau put it — evolution's creation — properly ought to be our goal. Once it was here, then we trampled it to destruction almost without a sidelong glance. Ben Goldfarb's *Eager* helps show us how to get it back.

— Dan Flores
Galisteo Valley, New Mexico

Introduction

The first time I tried to meet Drew Reed, the most prolific beaver mover in the state of Wyoming, I was thwarted by a sick goat. Reed and I had made plans to rendezvous in Jackson, the tony resort town south of Grand Teton and Yellowstone National Parks; I was en route when Reed called, his drawl pinched with concern. One of his goats, a 220-pounder named Maximus, had been laid low by a mysterious malady, and the vet needed to intervene ASAP. Reed was sorry, but we'd have to postpone until his beloved creature had been cured. Maybe, he added hopefully, Maximus just needed some electrolytes. I was disappointed, but also felt some admiration — here was a guy who cared enough about animals that he'd blow off a human engagement to tend to the health of a ruminant.

A month later, with Maximus in finer fettle, I found myself in the front seat of Reed's pickup, bouncing along a dirt road through the Gros Ventre Valley. A transverse crack glittered in the windshield; the rear window was plastered with a decal of a *T.-rex*-sized billy terrorizing tiny humans, accompanied by the phrase MY GOAT ATE YOUR STICK FAMILY. A boxy white trailer rattled in our wake, threatening to decouple from Reed's truck with every rut. Reed and his wife, Amy, normally used the trailer to tote Maximus. Today it conveyed a more sensitive cargo.

"Hope they're doing okay back there," Reed muttered.

The road, pocked with potholes deep enough to drown in, clung tight to the hillside, nearly forcing us to trade paint with cars creeping in the opposite direction. The Gros Ventre River ran below, a shimmering blue thread twisting through sere sagebrush meadows. Reed, a baseball hat yanked low over his shaved head, grumbled at unyielding drivers. At last the road descended into the valley, where an amber tributary called Cottonwood Creek gushed into the mainstem of the Gros Ventre. Reed executed a nimble three-point turn and backed his rig down to the creek so that the trailer's rear door opened onto the water. He clambered down from the cab for a quick debrief.

"The main thing we have to be prepared for is, they could separate," he warned me and a few onlookers who'd followed in a separate car. "It's doubtful they're gonna go upstream like I'd like for 'em to do — path of least resistance and all. I'm gonna do everything in my power to keep 'em together. The last thing we need is someone running between 'em." He paused for emphasis. "The welfare of these animals is *always* paramount over people's enjoyment."

A chorus of "I agree!" rose from the small crowd. "All right, then," Reed said. He unlatched the trailer door, lowering it into the creek to form a ramp. Then he stepped back.

The beaver who poked her head from the straw-covered bed of Drew Reed's trailer was *big* — big enough to make me inhale involuntarily. If you have ever seen a beaver, you have probably seen her swimming at a distance, with most of her estimable mass concealed, iceberg-like, underwater: a misleading view that creates the impression that beavers are little larger than housecats. Not so. This animal weighed sixty pounds, as much as many golden retrievers, a dense bolus of muscle and fat and milk chocolate fur — the linebacker of the animal kingdom. She — a pronoun I assigned her at random, as beavers' sexes are notoriously difficult to discern — stood precariously on her hind legs in the doorway, nose twitching as she surveyed her surroundings, front paws held to her chest tentatively like Oliver Twist asking for more gruel. But her caution didn't last long: Here was running water and standing cottonwood, all the habitat and food that a bark-noshing semi-aquatic rodent could desire. The beaver dropped to all fours and waddled down the ramp, hips and rump swaying like the ponderous bulk of a stegosaurus. This was not an animal well suited for land travel.

"Hey, bud!" Reed cooed. "Water, huh? You like water?"

No sooner had the hefty adult emerged than she was followed by a baby beaver, a kit, hardly bigger than a Chihuahua. We murmured our delight; even the hard-boiled Reed, I figured, would have to admit the thing was pretty dang cute. The juvenile hesitated, and Reed gave it a swift pat on the butt, as you would to an obstinate horse. "Go with Mom," he chided. The two beavers scudded into the stream, weaving back and forth, half swimming and half walking, the water not quite deep enough to submerge.

They looked understandably disoriented — they'd endured a long journey in a dark chamber, been flung into new environs, and were surrounded by strange hairless bipeds. Their ordeal, I thought, was like getting snatched by aliens from your bed in Sacramento, spending a day in isolation aboard a mysterious mothership, and then being dumped unceremoniously into a cornfield in Topeka.

That confusion, perhaps, explained what happened next. With a flick of his oar-like tail, the kit abandoned his parent and took off downstream, slipping like a trout over a rocky rapid. In defiance of Reed's fervent wishes, the pair was separating. Absent an adult, the kit would surely perish, either of starvation or in a cougar's jaws. Reed dashed over the cobble toward the bottom of the rapid, where he stood in shin-deep water, crouched like a shortstop preparing to corral a wicked grounder. Deftly he plunged his arm into the stream and, to our astonishment, hoisted the kit up by his leathery tail, holding him aloft like a trophy fish. Other handlers, before and since, have warned me against carrying beavers by the tail, for fear of dislocating the appendage. Although Reed isn't persuaded by the dislocation theory, he doesn't make a habit of tail grabbing, either. But in the heat of the moment, what choice did he have?

"That little booger got into that deep hole right in front of me and I was like, oh crap!" Reed told me later, after the beavers had been reunited and shooed upstream. "There's really no other place on a beaver to grab 'em."

The Gros Ventre River flows into Jackson Hole, the glacier-flattened valley that lies beneath the bladed Teton Range. Today Jackson Hole is a playground for the Patagonia class, a ritzy sprawl of ski slopes and mountain biking trails and upscale art galleries. Two centuries ago, though, the valley was defined by fur. In the autumn of 1807, John Colter, a former member of the Lewis and Clark expedition, followed the Bighorn River into the Rocky Mountains to trade with the Crow Indians. Colter wandered Wyoming for months in the snowbound dead of winter, toting little more than a rifle and a pack. Although no one's quite certain where his route took him, he's considered the first white man to enter the hole, a word trappers used to describe broad, game-filled valleys. He also found lots of beavers.

In the decades that followed, a parade of fortune seekers followed Colter's footsteps into the Northern Rockies, a region that, blared one newspaper, "possess[ed] a wealth of furs not surpassed by the mines of Peru."[1] These travelers were the famed mountain men, rapacious beaver trappers who, between the early 1820s and the late 1840s, systematically ransacked just about every pond and stream between Colorado and California. Most of those pelts flowed to the Missouri River and thence to St. Louis, to be shipped off to the East Coast or Europe for conversion into fashionable hats. With breathtaking speed, the mountain men demolished their resource, virtually wiping out beavers throughout the American West. "The trappers often remarked to each other as they rode over these lonely plains that it was time for the white man to leave the mountains," Osborne Russell, a beaver hunter who frequented Wyoming and Utah, wrote in 1841, "as beaver and game had nearly disappeared."[2]

Although the reign of the mountain men was brief, they left an enduring ecological legacy. If you know nothing else about beavers, you're probably aware that they build dams: walls of wood, mud, and rock that hold back water and form ponds and wetlands. The rodents also construct lodges, towering houses that often rise from open water like volcanic islands. These structures don't just house beavers themselves: Trumpeter swans squat rent-free atop beaver lodges, commandeering them as nesting platforms upon which their chicks shelter from land-bound predators like foxes. The majestic white birds also crave the elodea, sago pondweed, and other aquatic plants that grow in shallow beaver ponds.

By trapping out the Northern Rockies' beavers, the mountain men unwittingly destroyed countless acres of prime swan habitat. A few decades later farmers and ranchers finished the job by draining wetlands to make way for cattle and alfalfa. Today only ninety or so resident trumpeter pairs linger in the region, and chicks seldom survive. "Beaver ponds would've been strung out like necklaces down these drainages, and this landscape would have been a giant sponge," a swan biologist named Ruth Shea told me. "That's why there were swans nesting everywhere. Swans are the poster child for the importance of the beaver."

By the dawn of the twentieth century, the fur trade had largely dissolved, a victim of its own success. Beavers began to recover, much to the chagrin of Jackson's landowners, who rang up wildlife control trappers whenever the rodents gnawed down cottonwoods, dammed irrigation ditches, or flooded fields. No longer did we regard beavers primarily as pelts — just pests.

That didn't sit right with Drew Reed, an Arkansas native who, in 2008, took a job at the Wyoming Wetlands Society. Intrigued by beavers' ecological potential, Reed set out to make capture-and-relocation a priority. He taught himself to live-trap and hung up flyers advertising his services. Word of Reed's humane approach spread among Jackson's wildlife-loving citizenry like brushfire in dry grass. "All of a sudden my phone was ringing off the hook," he told me. Some trappers threatened their new competitor; others referred him clients. Before long he was dumping beavers in the Gros Ventre River two or three times a week. Filmmakers arrived from the BBC to shoot a documentary, salaciously titled *Beavers Behaving Badly*.

In 2015 Reed and Shea scraped together funding for a new nonprofit, the Northern Rockies Trumpeter Swan Stewards. Birds may be their remit, but their focus is beavers. Usually Reed arrests his quarry in suitcase-like live traps, though sometimes he's forced to get creative. Just before I came to Jackson Hole, he wrangled an especially wily fugitive with a salmon net — "a harebrained scheme," he gleefully acknowledged. "It was utter chaos when that beaver hit the net. It was a rodeo." He held on. The beaver was moved.

All told, Reed estimates he's relocated north of 250 beavers. How many have survived is another question. Although he's recaptured some old friends years later, many, no doubt, have been devoured by bears, wolves, and cougars, or slain by trappers. Without Reed, though, their fate would have been more certain, and grimmer. "Even if one landowner is willing to let the beavers stay, their neighbors probably won't — and we all know that beavers don't understand what a property line is," he told me as we bounced home. "I'm usually given an ultimatum: You relocate 'em, or they're dead. We're giving them a second lease on life, a chance to try to go make it. I call it reseeding a drainage. They're not gonna stay exactly where you put 'em, but I'm happy if they stay somewhere in the area and start doing their work."

As if on cue, Reed threw the truck in park and raised binoculars, ogling a lodge protruding from a distant pond. The structure, he told me, was likely the handiwork of a relocatee: He'd recently spotted one of his ear-tagged beavers cruising around the complex. "Oh, heck yeah — that thing has *grown*," he enthused. "That's three times the size it was a couple years ago." He gazed into the floodplain below, the wide sagebrush pasture parched and sepia and swanless as it rolled away from the river. "Man," he murmured, almost to himself, "I'd love to see that whole meadow underwater."

———

Close your eyes. Picture, if you will, a healthy stream. What comes to mind? Perhaps you've conjured a crystalline, fast-moving creek, bounding merrily over rocks, its course narrow and shallow enough that you could leap or wade across the channel. If, like me, you are a fly fisherman, you might add a cheerful, knee-deep angler, casting for trout in a limpid riffle.

It's a lovely picture, fit for an Orvis catalog. It's also wrong.

Let's try again. This time, I want you to perform a more difficult imaginative feat. Instead of envisioning a present-day stream, I want you to reach into the past — before the mountain men, before the Pilgrims, before Hudson and Champlain and the other horsemen of the furpocalypse, all the way back to the 1500s. I want you to imagine the streams that existed before global capitalism purged a continent of its dam-building, water-storing, wetland-creating engineers. I want you to imagine a landscape with its full complement of beavers.

What do you see this time? No longer is our stream a pellucid, narrow, racing trickle. Instead it's a sluggish, murky swamp, backed up several acres by a messy concatenation of woody dams. Gnawed stumps ring the marsh like punji sticks; dead and dying trees stand aslant in the chest-deep pond. When you step into the water, you feel not rocks underfoot but sludge. The musty stink of decomposition wafts into your nostrils. If there's a fisherman here, he's thrashing angrily in the willows, his fly caught in a tree.

Although this beavery tableau isn't going to appear in any *Field & Stream* spreads, it's in many cases a more historically accurate picture — and, in crucial ways, a much healthier one. In the intermountain West, wetlands, though they make up just 2 percent of total land area, support 80 percent of biodiversity; you may not hear the tinkle of running water in our swamp, but listen closely for the songs of warblers and flycatchers perched in creekside willows. Frogs croak along the pond's marshy aprons; otters chase trout through the submerged branches of downed trees, a forest inverted. The deep water and the close vegetation make the fishing tough, sure, but abundant trout shelter in the meandering side channels and cold depths. In *A River Runs Through It*, Norman Maclean captured the trials and ecstasies of angling in beaver country when he wrote of one character, "So off he went happily to wade in ooze and to get throttled by brush and to fall through loose piles of sticks called beaver dams and to end up with a wreath of seaweed round his neck and a basket full of fish."[3]

And it's not just fishermen and wildlife who benefit. The weight of the pond presses water deep into the ground, recharging aquifers for use by downstream farms and ranches. Sediment and pollutants filter out in the slackwaters, cleansing flows. Floods dissipate in the ponds; wildfires hiss out in wet meadows. Wetlands capture and store spring rain and snowmelt, releasing water in delayed pulses that sustain crops through the dry summer. A report released by a consulting firm in 2011 estimated that restoring beavers to a single river basin, Utah's Escalante, would provide tens of millions of dollars in benefits each year.[4] Although you can argue with the wisdom of slapping a dollar value on nature, there's no denying that these are some seriously important critters.

To society, though, beavers still appear more menacing than munificent. In 2013 I lived with my partner, Elise, in a farming town called Paonia, set high in the mesas of Colorado's Western Slope. Our neighbors' farms and orchards were watered by labyrinthine irrigation ditches, each one paralleled by a trail along which the ditch rider — the worker who maintained the system — drove his ATV during inspections. In the evenings we strolled the ditches, our soundtrack the faint gurgle of water through headgates, our backdrop the rosy sunset on Mount Lamborn. One dusk we spotted a black head drifting down the canal like a piece of floating timber. The beaver let us approach within a few feet before slapping his tail explosively and submarining off into the crepuscule. On subsequent walks we saw our ditch beaver again, and again, perhaps half a dozen times altogether. We came to expect him, and though it was probably our imaginations, he seemed to grow less skittish with each encounter.

Like many torrid romances, our relationship acquired a certain frisson from the knowledge that it was doomed. Although our beaver showed no inclination to dam the canal — and indeed, beavers often elect not to dam at all — we knew the ditch rider would not tolerate the possibility of sabotage. The next time the rider passed us on his ATV, a shotgun lay across his knees. The grapevine gave us unhappy tidings a few days later: Our ditch beaver was no more.

That zero-tolerance mentality remains more rule than exception: Beavers are still *rodenta non grata* across much of the United States. They are creative in their mischief. In 2013 residents of Taos, New Mexico, lost cell phone and internet service for twenty hours when a beaver gnawed through a fiber-optic cable.[5] They have been accused of dropping trees atop cars on Prince Edward Island,[6] sabotaging weddings in Saskatchewan,[7] and ruining golf courses in Alabama — where, gruesomely, they were slaughtered with pitchforks, a massacre one local reporter called a "dystopian *Caddyshack*."[8] Sometimes they're framed for crimes they did not commit: Beavers were accused of, and exonerated for, flooding a film set in Wales.[9] (The actual culprits were the only organisms more heedless of property than beavers: teenagers.) Often, though, they're guilty as charged. In 2016 a rogue beaver was apprehended by authorities in Charlotte Hall, Maryland, after barging into a department store and rifling through its plastic-wrapped Christmas trees.[10] The vandal was shipped off to a wildlife rehab center, but his comrades tend not to be so lucky.

Although our hostility toward beavers is most obviously predicated on their penchant for property damage, I suspect there's also a deeper aversion at work. We humans are fanatical, orderly micromanagers of the natural world: We like our crops planted in parallel furrows, our dams poured with smooth concrete, our rivers straitjacketed and obedient. Beavers, meanwhile, create apparent chaos: jumbles of downed trees, riotous streamside vegetation, creeks that jump their banks with abandon. What looks to us like disorder, though, is more properly described as complexity, a profusion of life-supporting habitats that benefit nearly everything that crawls, walks, flies, and swims in North America and Europe. "A beaver pond is more than a body of water supporting the needs of a group of beavers," wrote James B. Trefethen in 1975, "but the epicenter of a whole dynamic ecosystem."[11]

Beavers are also at the center of our own story. Practically since humans first dispersed across North America via the Bering Land Bridge — replicating a journey that beavers made repeatedly millions of years prior — the rodents have featured in the religions, cultures, and diets of indigenous peoples from the nations of the Iroquois to the Tlingit of the Pacific Northwest. More recently, and destructively, it was the pursuit of beaver pelts that helped lure white people to the New World and westward across it. The fur trade sustained the Pilgrims, dragged Lewis and Clark up the

Missouri, and exposed tens of thousands of native people to smallpox. The saga of beavers isn't just the tale of a charismatic mammal — it's the story of modern civilization, in all its grandeur and folly.

Despite the fur trade's ravages, beavers today face no danger of extinction: Somewhere around fifteen million survive in North America, though no one knows the number for certain. In fact, they're one of our most triumphant wildlife success stories. Beavers have rebounded more than a hundredfold since trappers reduced their numbers to around one hundred thousand by the turn of the twentieth century. The comeback has been even more dramatic across the Atlantic, where populations of a close cousin, the Eurasian beaver (*Castor fiber*), have skyrocketed from just one thousand to around one million.[12] Not only have beavers benefited from conservation laws, they've helped author them. It was the collapse of the beaver — along with the disappearance of other persecuted animals, like the bison and the passenger pigeon — that sparked the modern conservation movement.

But let's not pat ourselves on the backs too heartily. As far as we've come, beaver restoration has many miles farther to go. When Europeans arrived in North America, the naturalist Ernest Thompson Seton guessed that anywhere from sixty million to four hundred million beavers swam its rivers and ponds.[13] Although Seton's appraisal was more than a bit arbitrary, there's no doubt that North American beaver populations remain a fraction of their historic levels. Will Harling, director of the Mid Klamath Fisheries Council, told me that some California watersheds host just one one-thousandth as many beavers as existed before trappers pursued them to the brink of oblivion.

That story, of course, isn't unique to California, or to beavers. Europeans began despoiling North American ecosystems the moment they set boots on the stony shore of the New World. You're probably familiar with most of the colonists' original environmental sins: They wielded an ax against every tree, lowered a net to catch every fish, turned livestock onto every pasture, churned the prairie to dust. In California's Sierra Nevada, nineteenth-century gold miners displaced so much sediment that the sludge could have filled the Panama Canal eight times.[14] We are not accustomed to discussing the fur trade in the same breath as those earth-changing industries, but perhaps we should. The disappearance of beavers dried up wetlands and meadows, hastened erosion, altered the course of countless streams, and imperiled water-loving fish, fowl, and amphibians — an aquatic Dust Bowl. Centuries before the Glen Canyon Dam plugged up the Colorado and the Cuyahoga burst into flame, fur trappers were razing stream ecosystems. The systematic extirpation of beavers, wrote Suzanne Fouty in 2008, represented "the first large-scale Euro-American alteration of watersheds."[15]

If trapping out beavers ranked among humanity's earliest crimes against nature, bringing them back is a way to pay reparations. Beavers, the animal that doubles as an ecosystem, are ecological and hydrological Swiss Army knives, capable, in the right circumstances, of tackling just about any landscape-scale problem you might confront. Trying to mitigate floods or improve water quality? There's a beaver for that. Hoping to capture more water for agriculture in the face of climate change? Add a beaver. Concerned about sedimentation, salmon populations, wildfire? Take two families of beaver and check back in a year.

If that all sounds hyperbolic to you, well, I'm going to spend this book trying to change your mind.

———

Like most people who enjoy mucking about in streams, I've had my share of beaver encounters. I was always impressed by their underwater grace, their ingenuity, and their familial devotion; I once watched a mated pair fastidiously groom each other for a solid half hour in Glacier National Park. But I didn't become a true acolyte until a dreary Seattle morning in January 2015, when I shook off the damp and pushed into a fluorescent-lit Marriott conference room.

It was an unlikely setting for a profound conversion, but enlightenment is not known to strike predictably. Over eight hours a parade of tribal, federal, and university scientists — clad almost exclusively in flannel, the customary uniform of northwestern biologists — presented compelling evidence that our landscapes had been blighted by an absence of beavers, and that bringing them back was the most effective way to right a host of wrongs. The conference revealed a hidden world underlying my staid perception of aquatic ecology, one in which a pudgy rodent was responsible for everything from the dimensions of southwestern arroyos to Oregon's prolific salmon runs. Like the diligent journalist that I am, I had forgotten my notebook. Instead I scribbled down thoughts on cocktail napkins; by day's end my pile was covered in exclamation points and capital letters: FLOODPLAIN CONNECTIVITY! SLOW-WATER REFUGIA! RIPARIAN VEGETATION FEEDBACK LOOP! I walked in a feckless agnostic; I walked out a disciple.

That summer, still captivated, I traveled to central Washington to visit one of the West's foremost beaver evangelizers: Kent Woodruff, then a Forest Service biologist and the director of the Methow Beaver Project. For three days Woodruff interpreted his beaver-influenced corner of the country and brokered a personal introduction to the animals themselves. (*Very* personal: By my second morning Woodruff had me grappling with a hefty male in an attempt to harvest some of his anal secretions — the beaver's, that is.) The articles I wrote about the Methow Project's approach to beaver relocation (see chapter 4 for much more) appeared in the magazine *High Country News*, and ultimately birthed this book.

In researching *Eager*, I traveled just about everywhere that beavers can be found, from the slickrock deserts of Utah to the hardwood forests of Vermont to a highwayside canal in Napa, California. I met beavers on farms and beavers in forests, beavers in raging rivers and beavers in irrigation ditches, beavers in wilderness areas and beavers in Walmart parking lots. Nor did I confine myself to North America: Elise and I also journeyed to the moors of southwest England and the sheep-dappled hills of the Scottish Highlands to document beavers' fitful return to Britain. Although beavers could historically be found just about anywhere in the United States — excepting south Florida, where they would make hors d'oeuvres for alligators — this book is primarily set west of the 100th meridian, the line of longitude that carves through the Great Plains. Beyond that decisive boundary, rainfall dries up and beavers become even more crucial. "Back east you have water no matter what, but we have streams that run dry," Mary O'Brien, a scientist based in arid southeast Utah, told me when I visited. "And beavers can just make wetlands appear here. They're kind of magic."

O'Brien and her ilk form nothing short of a movement, a growing coalition of wildlife biologists and land managers and renegade ranchers who are protecting and restoring beavers for every reason under the sun — to create butterfly habitat, to nourish cattle, to purify drinking water, to rebuild eroded river valleys. And the adherents of this movement have a name: They call themselves Beaver Believers.

There is no single trait that unites Beaver Believers, besides, of course, the unshakable conviction that our salvation lies in a rodent. (They also share the tendency to proselytize: *Shy* is not a word you'd use to describe the movement's members.) In an era when most environmental actions spawn polarized outrage, Beaver Believers span party lines: You'll meet plenty of dyed-in-the-wool beaver huggers in this book, but you'll also encounter red state stockmen. Many devotees are trained biologists. Many, I've noticed, are not. The world's most knowledgeable Beaver Believers include former hairdressers, physician's assistants, chemists, and child psychologists. Perhaps there are dozens of ex-laypeople fighting for weasels and kangaroo rats, but somehow I doubt it. There is, I think, something uniquely beguiling about beavers: their ability to support other species, their complex and endlessly interpretable behavior, their fundamentally human attitude toward modifying landscapes. They are *visible* in a way that few other species are — and so are their admirers. "Now people see me in town and say, 'Oh, you're the beaver gal!'" Charnna Gilmore, a real estate agent turned devout Believer, told me when we met in California's Scott Valley. She grinned defiantly, the smile of one who has found inner peace. "My family, they think I'm having a midlife crisis."

People outside the cult, I've found, tend not to share Gilmore's affection for our tree-chewing brethren. Since my conversion to Believerdom, I have rhapsodized about the virtues of beavers to more than my fair share of friends, family, and complete strangers at bars signaling frantically for their checks. Usually I get polite laughter and dirty jokes. (Admit it — you were thinking it.) Sometimes I get stories: about fathers losing their prized apple trees to rascally beavers, tail-slaps shattering the stillness of an Adirondack lake, childhood hours whiled away dynamiting beaver dams on a Montana ranch. People often tell me they think beavers are cute.

And that's wonderful: I think beavers are cute, too. But I urge you, dear reader, not to underestimate these extraordinary mammals. Many animals are cute; very few are ecosystem engineers. Even acknowledging that beavers store water and sustain other creatures is insufficient. Because the truth is that beavers are nothing less than continent-scale forces of nature, in large part responsible for sculpting the land upon which we Americans built our towns and raised our food. Beavers shaped North America's ecosystems, its human history, its geology. They whittled our world, and they could again – if, that is, we learn to treat them as allies instead of adversaries. Our future must be as entwined with beavers as our past has been, and yet we must completely reverse the nature of our relationship. They will build it, if we let them come.

As Melville wrote, "To produce a mighty book, you must choose a mighty theme."[16] The book you hold in your hands may not be *Moby-Dick*, but it's not the theme's fault. The story of beavers is the story of how North America was colonized; why our landscapes look the way they do and how they've changed; what measures we can take to forestall the deterioration of our rivers, the disappearance of our biodiversity, and the ravages of climate change. Most of all, *Eager* is about the mightiest theme I know: how we can learn to coexist and thrive alongside our fellow travelers on this planet.

THE NORTH AMERICAN BEAVER

(Castor canadensis)

LEGEND

▨ Modern range of *Castor canadensis*

SELECTED SITES VISITED BY AUTHOR

1. Gros Ventre Valley, WY
2. Agate Fossil Beds
   National Monument, NE
3. Centennial Valley, MT
4. Fort Bragg, NC
5. Grafton, VT
6. Westford, MA
7. Methow Valley, WA
8. Bridge Creek, OR
9. Scott Valley, CA
10. Martinez, CA
11. Occidental Arts
    and Ecology Center, CA
12. Elko, NV
13. Yellowstone National Park
14. Logan, UT
15. Moab, UT
16. Bronx River, NY

— CHAPTER ONE —

Appetite for Construction

To be human is to be a survivor. *Homo sapiens* are the world's only living hominids, and we've likely been alone for around the past forty thousand years. But our present solitude is a recent development. For millennia we shared this planet with bipedal cousins: Neanderthals roamed the forests and beaches of Europe, the Denisovans wandered Southeast Asia, the hobbit-like *Homo floresiensis* sheltered in Indonesian caves. Why we pulled through while our cousins perished remains somewhat mysterious, but we probably triumphed through some combination of innovative tool use and demographic luck. In that regard, we have company.

In 1891 a geologist named Erwin Hinckley Barbour was called to examine a fossil. Barbour, an angular man whose mustache, at full wax, outflanked his face, boasted an impeccable scientific pedigree: At Yale he'd studied under O. C. Marsh, the world's first professor of paleontology. During his half-century-long career, Barbour would become Nebraska's state geologist and show a particular affinity for extinct elephants, describing more than a dozen vanished mastodons, mammoths, and four-tusked gomphotheres. If you needed a midwestern fossil identified in the late nineteenth century, Barbour was your man. Yet when a Nebraska rancher named James Cook requested that Barbour lend his talents to a huge sandstone spiral he'd discovered on his property, not even the august scientist had the faintest idea what he was looking at.

Although local ranchers were no strangers to the confounding spirals, the fossil that confronted Barbour resembled nothing he'd ever seen.

Clearly it was no bone. The specimen was a vertical stone helix, taller than a man, that resembled "a great three-inch vine coiled . . . about a four or five inch pole," like a strangler fig wrapped around a sapling. Some adjacent helices, Barbour estimated, "could not be less than 30 or more feet in height." Each pole terminated in a "transverse piece," a thicker, perpendicular base, like the handle of an upside-down corkscrew. The spirals peppered several square miles of Nebraska badlands. "These fossils seem altogether so remarkable and of such imposing size and peculiarity of form," Barbour wrote, "that I have felt great hesitancy into offering any suggestions as to what they are."[1] The flummoxed geologist hazarded that he'd discovered a gargantuan freshwater sponge; later, he revised his diagnosis to the root castings of a huge plant. Neither guess proved correct. Still, the name Barbour gave the spirals stuck. He called them *daemonelices*, a highfalutin Latinization of the ranchers' old moniker: Devil's corkscrews.

In investigating the cryptic corkscrews, Barbour noticed that the helices were lined with vegetal material and, on occasion, rodent bones, which he interpreted as evidence of his giant plant theory. When the Austrian paleontologist Theodor Fuchs analyzed the daemonelices in 1893, however, he realized the structures weren't organisms at all — rather, they were titanic burrows carved into the earth by the rodents entombed at their bottom. In 1905 Olaf A. Peterson finally described the diggers, identifying them as members of a genus that would come to be known as *Palaeocastor*: literally, "ancient beaver."[2]

At least superficially, *Palaeocastor*, in its subterranean habits and small stature, less resembled contemporary beavers than it did gophers. Although they possessed hefty front teeth, the proto-beavers didn't use their incisors to fell trees or wood. Instead they were dental excavators. "The walls [of the burrows] were covered with broad grooves that I could match by scraping the incisors of the fossilized beaver skulls into wet sand," Larry Martin, a paleontologist who examined more than a thousand daemonelices during the 1970s, reported in *Natural History*. "The beavers had used their teeth to scrape dirt off the walls. . . . A burrowing beaver must have fixed its hind feet on the axis of the spiral and literally screwed itself straight down into the ground" — not unlike a drill bit twisting into a board.[3] They'd cleared away loose sand not by flinging it with their claws, but by pushing it out with their broad, flat heads.

If all of that strikes you as a gritty way to eke out a living, bear in mind that *Palaeocastor* survived for about four times longer than *Homo*'s run to date. The spirals may have allowed the beavers to cluster many deep burrows in a small area; deterred predators; modulated humidity and temperature; or helped the animals survive flooding. But not even their unique abodes saved ancient beavers from extreme planetary makeover. As temperatures fell, the world dried up; *Palaeocastor*, adapted for a wetter climate, went extinct during the early Miocene, around twenty million years ago. The beavers who dug Barbour's daemonelices were among the last of their lineage, give or take a few million years.

Although hundreds of corkscrews remain scattered around the badlands, nearly all are sequestered on private ranches. Your best option for visiting the Miocene today is to trek to Nebraska's Agate Fossil Beds National Monument, a charming, middle-of-nowhere parcel of limestone-capped buttes and lush hollows that receives fewer tourists in a year than Yellowstone gets on a busy summer day. One June morning I parked near the monument's entrance and walked up a short trail toward a sculpted sandstone bluff that towered above the prairie like a tallship. Twenty million years ago Nebraska resembled nothing so much as the Serengeti, a river-webbed grassland upon which foraged a spectacular mammalian bestiary: tiny camels and giant wolverines, two-horned rhinos and pig-like oreodonts, muscular beardogs and lithe horses. Since the Miocene the land has gained contour and lost wildlife, though a sign at the trailhead advised me to avoid rattlesnakes. Trailside placards identified evocatively named prairie plants: downy paintbrush, bottlebrush squirreltail, rubberweed. I could have driven a golf ball in any direction without striking a tree.

I found the daemonelix at the base of a gray outcropping, its curves etched in sandstone like Han Solo frozen in carbonite. The towering spiral stood behind a smudged glass case, its corkscrew geometry so close to perfect that it looked unnatural. I sat on a nearby bench, listening to the burbling meadowlarks and whirring grasshoppers, feeling like a religious pilgrim come to genuflect before a sacred relic. Here it was — among the holiest objects in Beaverdom, a castorid mecca.

I hustled down the trail to catch up with a park ranger, his broad-brimmed hat bobbing over the prairie as he led a couple on a guided stroll. The ranger, a genial geologist named Trevor Williams, was explaining the competing hypotheses about the helix's function. His preferred theory was that the slight incline of the burrow's bottom chamber offered a refuge against the flash floods that must have swept the ancient pan-flat plain.

"Kind of like how I try to drown my moles," the woman said.

Although the burrows had perplexed paleontologists, Williams said, local indigenous people had no misconceptions about the structures' architects. The spirals featured in the mythology of the Lakota, a tribe who say that ancient Thunder Beings turned beavers to stone to provide protection against devastating Water Monsters. It's impossible not to be amazed by these indigenous archaeologists: Even though the burrows in no way resemble modern beaver lodges, the Lakota weren't fooled. Their word for the rock helices was *Ca'pa el ti*, or "beavers' lodges" — far more accurate nomenclature than anything white people concocted.[4]

"I like to imagine Dr. Barbour presenting his tree root hypothesis to a room full of academic bigwigs, and everyone's clapping and saying, 'Bravo! Bravo!'" Williams said. "And the two Lakota guys are standing in the back of the room with their arms crossed, going 'Should we tell him about the beavers? Nah.'"

We walked down the short loop trail, and I thought about *Palaeocastor*'s world. What selective pressures drove its relatives to adopt an aquatic life-style? There's always a bit of conjectural storytelling involved in re-creating evolution, but they're still questions worth asking: Why *Castor canadensis* but not *Palaeocastor*; why *Homo sapiens* but not *Homo neanderthalensis*? What did our respective species figure out?

Williams, evidently, was pondering the same questions. "It's amazing to think that these guys are still around in some form," he said. "The beardog has no living relatives. The rhinos have disappeared from America. Of all the animals that used to live here, beavers are the big winner."

———

Although the evolutionary paths of rodents and primates forked more than eighty million years ago, don't let our divergent lineages fool you: Beavers are among our closest ecological and technological kin. *Homo sapiens* and *Castor canadensis* are both wildly creative tool users who settle near water, share a fondness for elaborate infrastructure, and favor fertile valley bottoms carved by low-gradient rivers. And while all organisms have evolved to fill niches provided by nature, neither beavers nor people are content to leave it at that. Instead we're proactive, relentlessly driven to rearrange our environments to maximize its provision of food and shelter. We are not just the evolutionary products of our habitat: We are its producers. If humans are the world's most influential mammals, beavers have a fair claim at second place.

Castoridae, the beaver family, evolved from the soup of Rodentia between thirty-five and forty million years ago, as tropical forests ceded to grasslands in the late Eocene. *Palaeocastor* may have been the oddest early beaver, but it wasn't the first; that honor likely goes to a little-known ancestor called *Agnotocastor*, a groundhog-like creature that probably sported a rat's scaly tail. (Their fossils are identifiable as early beavers by the shapes of their skulls.) Over millions of years around thirty genera of beavers evolved and vanished, from tiny blind root eaters to the hippo-like *Castoroides*, a beaver the size of a small black bear that roamed from Florida to Alaska and disappeared just ten thousand years ago. According to Pocumtuck legend, hills near Deerfield, Massachusetts, were molded from the bodies of giant beavers, perhaps a cultural memory from when our two species shared the earth.

All that remains of this once-flowering tree today is the genus *Castor*, comprising the familiar *Castor canadensis* in North America and *Castor fiber*, the Eurasian beaver, across the Atlantic. These two creatures, and the behaviors that we consider intrinsic to beaverhood — living in water and building dams — appear to be the scions of a mysterious progenitor whose novel adaptations shaped not only beavers' evolution, but the contours of two continents. While *Palaeocastor*'s side of the family tunneled into the Great Plains, the direct antecedents of modern beavers took to water.

Ellesmere Island is a windswept hunk of land, around the size of South Dakota, in the Nunavut territory, Canada's northernmost district. Save for the occasional willow, the island is layered in tundra, and patrolled by wolves and musk ox. It is no place for a beaver to live. Once, though, Ellesmere boasted a more forgiving climate and forests that resembled those you'd find in contemporary Montana. Some of those spruces and pines, alongside the bones of wolverines, deerlets, and horses, eventually sank into peat layers, thicker than a man is tall, at the bottom of an Arctic pond. That pond was also home to beavers — the source of some of the world's most important castorid fossils.

The beavers of Ellesmere's pond were members of a now-extinct genus called *Dipoides*. Like today's beavers it lived in water and chewed wood: Scientists have extracted *Dipoides*-gnawed sticks from the Ellesmere Beaver Pond, including a jumble of trunks and cobble that resemble the remnants of a dam. Although *Dipoides* was only two-thirds *Castor*'s size and couldn't match its gnawing ability, Natalia Rybczynski, the paleontologist who excavated the Ellesmere trove, wrote that it was nonetheless an "avid woodcutter" whose chewed sticks hold boggling significance.[5] Since it's unlikely that a behavior as bizarre as tree harvesting evolved more than once, the odds are good that the most recent ancestor shared by *Dipoides* and *Castor*, which lived around twenty-four million years ago, was also a wood-chewing, dam-building engineer. Why does that matter? As the Canadian author Frances Backhouse explained in her book *Once They Were Hats*, "The longer that beaver dams have been around, the more liable they are to have affected the evolution of a multitude of species, from aquatic invertebrates and plants to fish, amphibians, and wetland-dependent birds and mammals."[6] As we'll soon see, a menagerie of North American plants and animals rely upon beaver-created water features — and, if *Dipoides* was indeed a dam builder, that dependency may reach back twenty-four million years.

So who was the ur-cutter, the first beaver to use her incisors to gnaw off a stick — and to what end? In 2007 Rybczynski advanced two theories about the origins of wood cutting. By the early Miocene, the High Arctic had succumbed to climatic cooling, and lakes like the Ellesmere Beaver Pond had begun to freeze. Wood cutting may have helped early beavers construct "food caches": piles of edible sticks, driven into muddy pond bottoms, that modern beavers in northern climes still hoard to sustain themselves through hard winters. Or it may have allowed the rodents to build snug lodges, which would have been warmer than tunneling into the earth.[7]

From those humble beginnings, the conjoined behaviors of living in water and constructing woody architecture flourished. While the gopher-like burrowers died out, the aquatic side of the family wandered between North America and Eurasia via Beringia, the land bridge that once connected the continents, giving rise some ten million years ago to *Castor*, the genus that includes modern beavers. Although even experienced wildlife biologists struggle to tell North American and European beavers apart,

DNA analysis suggests their trajectories diverged around seven and a half million years ago, when intrepid Castor colonists emigrated back across the land bridge from Asia. After rising seas flooded Beringia and divided the continents two million years later, the lineages were isolated, free to independently evolve.[8] (If you're keeping track of all this, you'll notice that beavers arose in North America, crossed into Eurasia, and eventually returned, the prodigal rodent come home.) *Castor canadensis*, the modern North American beaver, arose more than a million years ago and, like the Clovis people who eventually followed, dispersed rapidly across a fertile continent. From the Alaskan interior to northern Mexico, Newfoundland to the Florida panhandle, where there was water, there were beavers. And where there were beavers, there was water.

The road to Taos Ski Valley in northern New Mexico follows the Rio Hondo, a dazzling mountain stream that trips through a steep defile cloaked in dark ponderosa pine. Near the banks the conifers cede to aspen, whose leaves, each fall, shiver golden on papery branches. Beavers dominate the Hondo: Every quarter mile the route winds past another limpid pool formed behind another stream-spanning dam. Taoseños have, for the most part, wisely resisted the temptation to build homes within this narrow, flood-prone valley. Drive far enough, though, and you'll come upon a scene of stunning watery conflict, where beavers and humans jousted for control of the valley bottom — and where humans surrendered.

I first visited the abandoned house on a September day with my friend Leah, a local who was aware of my beaver obsession. Around eight miles outside of Taos, we pulled off the road to admire the site. Calling it a "beaver pond" scarcely did it justice — "beaver industrial complex" seemed more apt. The resident rodents had laid down a hefty blockade of branches, some six feet tall and fifty feet long, in the middle of the Rio Hondo, diverting the creek into the front yard of a small red cottage with a peaked roof. The house stood derelict, a lonely island half submerged in a shallow lake. Black smears of water damage crept up the wall. The septic tank wallowed in a swamp. A few lonely telephone poles stood marooned, guy wires trailing uselessly into the pond. To our amazement, we saw that the beavers had deftly extended their main dam until it reached the front porch, whose supporting struts became, in turn, another piece of the immense wall. We wandered the marshy fringe, running our hands over chiseled aspen stumps. The few trees still standing, I noticed, were wrapped in chicken wire.

Although the property owners probably weren't too pleased, the compound was, to my eye, a magnificent feat of infrastructure. Each disparate design element functioned in harmony: The half-dozen adjunct dams, placed with surgical precision, shunted water into the spidery canals winding into the willows, an intricate network of channels that permitted the miniature Venetians to fell and transport trees without risking dangerous overland travel. Below the dam the river raced white-flecked and implacable; above, it spread serenely over grass and cobble, a half-acre harbor. More remarkable than the compound's immensity was its sophistication, the way the creatures had melded environment and architecture with an adroitness that recalled Frank Lloyd Wright. Leah, a relative beaver neophyte, was gobsmacked. "This is so impressive," she said, and I had to agree.

How do beavers engineer such epic works — and why go to such elaborate lengths? The primary reasons are the same ones that first drove humans to build domiciles of their own: safety from predators, shelter from the elements, and food storage. On land, beavers — North America's biggest rodent, and the world's second largest, after South American capybaras — are ungainly and vulnerable, and their pear-shaped bodies make delectable meals for black bears, cougars, coyotes, and wolves. Yet beavers are as balletic in water as they are clumsy out of it. They can hold their breath for up to fifteen minutes, and their underwater gymnastics are powered by webbed hind feet. Transparent eyelids allow them to see below the surface, while a second set of fur-lined lips close behind their teeth, permitting them to chew and drag wood without drowning. Building dams expands the extent of beavers' watery domains, submerges lodge entrances to repel predators, and gives them a place to stash their food caches. Ponds also serve to irrigate water-loving trees like willow, allowing beavers to operate as rotational farmers: They'll chew down vegetation in one corner of their compound while cultivating their next crop in another.

Like any smart construction workers, beavers begin their dams by laying the foundation, a low ridge of mud, stones, and sticks set perpendicular to the stream's flow. Next come long wooden poles, stacked at an angle and fixed in the bed, followed by smaller branches woven into the super-structure. (Beavers aren't picky about their materials: In 2016 canoeists found a prosthetic leg jammed into a dam in Forest County, Wisconsin. The prosthesis, fear not, was restored to its rightful owner, who'd offered a fifty-dollar reward on Craigslist for his limb's safe return.[9]) Finally, they caulk the gaps with mud, grass, and leaves. The structures come in an almost limitless range of shapes and sizes, from speedbumps the length of a human stride to a half-mile-long dike, visible from space, that winds through Albertan swampland. A single prolific colony, or family unit, can construct and maintain more than a dozen dams, converting a narrow stream into a broad chain of ponds. A typical compound of ponds and canals, writes the biologist Dietland Müller-Schwarze, serves as "highway, canal, lock (in two senses), escape route, hiding place, vegetable garden, food storage facility, refrigerator/freezer, water storage tank, bathtub, swimming pool, and water toilet."[10]

In addition to dams, beavers also create burrows and lodges, "both of which," one nineteenth-century observer wrote, "are indispensable to his security and happiness."[11] When convenient, the architects prefer the expediency of digging tunnels directly into a river- or lake-bank, just below waterline. Although *busy as a beaver* is a compliment in our labor-fetishizing society, the rodents are hardly monomaniacal builders; in the 1980s Russian researchers found that around a quarter of the beavers they studied were content to live inconspicuously in bank burrows.[12] When geography demands it, beavers construct more classic lodges: mounds of logs and sticks, sometimes as large as a human living room, set atop the bank or surrounded completely by water. Entrance tunnels lead from the snug darkness of an inner nest chamber, where the home builders sleep and rear kits, to the aqueous outside world. Beavers often plaster their fortresses with mud, which freezes into a concrete-hard sealant, rendering the lodge as winterized as an Earthship. Scientists in Minnesota discovered the cozy interiors hovered several degrees above freezing, even as temperatures out-side plummeted below zero degrees Fahrenheit.[13] Beavers don't hibernate, instead spending the winter dragging morsels from their submerged larder of sticks and roots to the family waiting at home. Trappers once watched for plumes of steam curling from a lodge's ventilation shaft to identify occupied huts, the beavers within betrayed by their own condensed breath.

To acquire their building materials, of course, beavers chew down trees — teetering precariously on hind legs, tail propped beneath the body like a kickstand, forepaws braced against trunk as they chip away with massive incisors. Once they've captured their ligneous prey, they dismantle it, gnawing off large limbs and sectioning unwieldy trunks for easier dragging. Beavers work as smart as they do hard: One Saskatchewan study found that 62 percent of felled trees toppled toward the dam, making it easier for the diminutive loggers to tug materials to their construction sites.[14] Still, accidents happen. In 1954 Harold Hitchcock, a biologist at Middlebury College in Vermont, reported a beaver crushed by a split ash tree, "its head pinched between the halves of the trunk."[15] Timber cutting is among the world's most dangerous professions, whether you fell hardwoods with chainsaws or teeth.

Stoop to examine the sticks that constitute a lodge or dam, and you'll usually find that they've been peeled smooth, as if whittled with a pocketknife. Beavers are reliably efficient, and, true to form, they generally eat the inner bark — the sugary tissue layer, known as the cambium, that does the growing — like corn on the cob before weaving sticks into their edifices. In summer, when the world is succulent, beavers graze on green plants as contentedly as any cow, munching everything from ferns to poison ivy; come winter, they switch to a woodier diet, subsisting on stems and bark. Their favorite foods include aspen, cottonwood, and willow, but they'll eat just about anything in a pinch. When one researcher sliced open beaver stomachs in Mississippi, he found them packed with forty-two tree species, thirty-six genera of green plants, four kinds of woody vines, and a sodden mulch of grasses.[16] Thrifty as ever, they also practice *caecotrophy*, eating their pudding-like excretions to extract every last iota of nutrition; by the time their feces reemerge a day later, they're nearly sawdust. Beavers manage to digest around a third of the cellulose they consume, a process that's aided not only by poop eating but also by an unusually long intestine and a spectacularly diverse microbiome.[17] In 2016 researchers found more than fourteen hundred species of bacteria dwelling in beaver feces — hundreds more than have been detected in our comparatively impoverished guts.[18]

Nonstop gnawing requires mighty dentition, and the beaver's famous chompers are up for the task. Beavers whittle with their incisors — two upper, two lower, all of which grow continuously to compensate for constant erosion. These peerless cutting tools are self-sharpening: The outer surface of beavers' front teeth — the side they show to the world — is coated in hard, dense enamel, while the reverse side is made of softer dentine. The inner face wears down more quickly than the outer, creating a beveled, chisel-like edge. The incisors are orange, exposing the iron that's built into the chemical structure of their enamel. Despite lacking toothbrushes and fluoride, beavers are remarkably resistant to tooth decay, a sensible adaptation for an animal that lives and dies by oral power tools.[19] If, however, a tooth jostles out of alignment, preventing it from filing down its counterpart, endless growth can become a hazard. Historical records are rife with reports of beavers who starved — and even allegedly impaled their own brains — after gruesome incisors grew amok.

Nearly as remarkable as beavers' teeth is their fur, the material so soft and pliable that it spurred the colonization of a continent. Beavers have two types of hair: coarse guard hairs, about two inches long, overlaying luxurious underfur, or wool. Their fur is thick, buoyant, and virtually waterproof, serving at once as armor, life preserver, and dry suit. Altogether, a stamp-sized patch of beaver skin is carpeted with up to 126,000 individual hairs — more than the average human has on her entire head. While the stiff guard hairs are plucked and discarded by hatmakers, the underfur is one of the finest materials in which humans have ever clothed themselves. In his poem "The Triumph," Ben Johnson analogized feminine beauty to swan's down, lilies, fresh snow, and "the wool of beaver."[20]

If Johnson had wanted to pay his paramour an even higher compliment, he might have compared her aroma to the rodents' glandular secretions. Beavers delineate their territories using their *castor sacs*, internal pouches that produce a musky, vanilla-tinged oil. The mammals, whose powerful noses make up for their mediocre eyesight, mix castoreum with urine and spray the cocktail onto mounds of dredged mud to warn off intruders, as if erecting pungent picket fences around property lines. In his *Natural History*, written in AD 77, the Roman naturalist Pliny the Elder describes castoreum as a miracle drug capable of curing headaches, loosening constipated bowels, and staving off epileptic fits. Fantastically valuable "beaver stones" later followed trade routes from western Europe to Persia and Africa, and doctors prescribed a dose for maladies including "gastralgia, dysentery, worms, retention of urine, induration of liver and spleen, pleurisy . . . gout, hysteria, hemicrania, and loss of memory."[21] In an anatomically preposterous fable dubbed "The Beaver and His Testicles," Aesop claimed that a beaver pursued by castoreum hunters would gnaw off its gonads and gift them to its tormentors. "As soon as the hunter lays his hands on that magical medicine, he abandons the chase and calls off his dogs," the Greek claimed. Never mind that castoreum comes from the castor sacs, not the testicles, and that a beaver's testicles are internal, so there's nothing to gnaw off.[22] Although castoreum's salutary properties were no doubt overblown, the craze wasn't entirely driven by junk science: Among dozens of other plant-based compounds, the substance contains salicylic acid, which the mammals derive from willow — and which happens to be the active ingredient in aspirin.

Aesop's testicular fable may have been the weirdest beaver-related myth, but it wasn't the only one. "Pictures have appeared . . . which show the houses with two stories, and with windows and doors cut *square*," groaned the naturalist Arthur Radclyffe Dugmore in his 1914 book *The Romance of the Beaver*. "It will not need much intelligence to see the absurdity of these 'facts.'"[23] The beaver's broad, pebbled, paddle-like tail provided another source of confusion. Fur trappers claimed beavers employed their tails as pile drivers to pound in sticks, and a print made in 1715 depicted a beaver assembly line using them as hods to lug rocks.

The reality is stranger than the legends. The beaver's flat, scaly tail is a nifty multi-tool exquisitely adapted to semi-aquatic life. In addition to serving as a kickstand, it's a rudder and an alarm system: If you've ever kayaked a beaver pond on a clear evening, you've likely had your serenity shattered by the gunshot of a tail striking water, a clarion signal that startles predators and sends nearby family members scrambling for safety. The tail is also lined with a *rete mirabile*, a "wonderful net" of tightly meshed blood vessels that exchange heat through their walls and regulate the beaver's temperature. Finally, the leathery appendage contains a substantial fat reserve that helps its owner endure hard winters. The creamy fat of a spit-roasted tail, wrote David Coyner, was a trapper's delicacy: "When the heat of the fire strikes through so as to roast it, large blisters rise on the surface, which are very easily removed. The tail is then perfectly white, and very delicious."[24]

Beavers are family-oriented creatures, and like many humans they're generally, though not exclusively, monogamous. A typical colony consists of four to ten members, including the mating adults; newborn kits, birthed

Beaver ponds, wetlands, and meadows are refuges not only for the rodents themselves, but for a menagerie of other plants and animals, including moose, otters, trumpeter swans, coho salmon, and, in North Carolina, the endangered Saint Francis' satyr butterfly. *Illustration by Sarah Gilman.*

in May or June; and yearlings born the previous spring. Two-year-olds tend to move out in search of their own territories soon after the birth of their youngest siblings, like teenagers heading off for college – though in Quebec, Françoise Patenaude observed two-year-olds remaining at home to groom, feed, and guard their kid brothers and sisters.[25] For kits, their two-year homestays serve as internships, during which they learn and hone tree-felling, dam construction, predator avoidance, and other aptitudes. The naturalist Hope Ryden, in her book *Lily Pond*, observed kits and yearlings picking up a wide variety of behaviors from their caretakers, from the finer points of food-caching to rolling up a lily pad, burrito-style, for easier munching.[26]

As much as beavers value nurture, nature still hardwires their most important behavioral building blocks. In the 1960s Lars Wilsson, a Swedish ethologist, performed a series of clever experiments on captive-born beavers who lacked dam-building experience. Wilsson found that the naive beavers, released into running water, erected exemplary dams on their first attempts. When Wilsson subsequently played the trickle of flowing water over a loudspeaker in a dry room, the confused creatures built dams across the concrete floor.[27]

Although beavers' damming instincts produce some laughable results in captivity, in the wild their legendary assiduousness is an asset. Beavers, wrote Arthur Radclyffe Dugmore, "seldom allow a night to pass . . . without making a tour of inspection and building up and strengthening any part that shows signs of weakness."[28] In "flashy" streams whose courses are annually scoured by huge pulses of snowmelt, dams are often ephemeral, washing out each spring and requiring reconstruction when the waters recede in summer. But when conditions allow, beaver compounds can prove astonishingly durable, maintained by succeeding generations for decades — or even centuries.

Much of what we know about the extraordinary longevity of beaver complexes comes courtesy of Lewis Henry Morgan. Morgan, an anthropologist and railroad profiteer, was a dubious scholar whose work often descended into racism — he notoriously classified human cultures as either savage, barbaric, or civilized — but an undeniably keen observer of castorids. His travels to Michigan's Upper Peninsula, where he sought to open an iron mine, brought him into close contact with the rodent, whose dams and canals soon caught his anthropological fancy. "There is no mammal, below man . . . which offers to our investigation such a series of works," Morgan enthused in his 1868 book, *The American Beaver and His Works*, "or presents such remarkable materials for the study and illustration of animal psychology."[29]

Morgan's corner of Michigan was then a beaver paradise, an "unbroken and an uninhabited wilderness" webbed with creeks and forested in birches, maples, and willows. By Morgan's count, beavers had constructed sixty-three dams in the area, including one colossus that stretched 260 feet long and stood over 6 feet high. "It has undoubtedly been built upon and repaired year after year until it reached its present dimensions; and it is not in the least improbable that it has existed and been continued for centuries," he wrote. Morgan gushed about the structures' "remarkably artistic appearance," their strategic placement, and their craftsmanship. "They appeared to be loosely thrown together, but on attempting to raise a number of them they were found to be fast at one end or the other, or so interlaced that it was difficult to remove them," he marveled. One earthen embankment, he wrote, was so solid and broad "that a horse and wagon might have been driven across the river upon it in safety."[30]

Morgan's dam observations were as precise as they were enthusiastic. With the help of two railroad engineers, he drafted a map of the region's rivers, railroads, and beaver-created lakes: Lake Helen, Grass Lake, Stafford Lake, and dozens more. Morgan's map is a work of delicate art, the kind of fanciful schematic you'd find at the outset of a Tolkien novel. It's also one of the most meticulous representations of a fully beavered landscape ever rendered. Although the region fell within the historic territory of the Ojibwe, for whom beaver served, as one chief put it, as "Indian's pork,"[31] the mining railroad represented virtually the first white intrusion into this damp region. The map folded into the 396-page *American Beaver* therefore provides a key to a lost world — and an invaluable tool for science.

More than a century after Morgan's death, Carol Johnston, an ecologist at South Dakota State University, learned about his map during the course of her postdoctoral research.[32] What, she wondered, had become of the beaver-built landforms that Morgan had documented? When she examined an array of contemporary aerial images in 2014, she discovered that three-quarters of the dams and ponds that appeared in *The American Beaver*'s diagram remained visible. Many of the sites appeared to have been abandoned, and, without active maintenance by their beavers, filled in with earth to become meadows or shrublands. Other ponds had actually grown in the past 150 years, an indication that beavers had raised their dams to capture still more water. Morgan's map and Johnston's update demonstrate the seeming paradox at the heart of beaver ecosystems: that even as topographic features evolve year by year, a colony's footprint can persist for decades or longer. Beaver landscapes are cauldrons of turmoil and bastions of stability, dynamic and durable in equal measure. "This constancy is evidence of the beaver's resilience," Johnston wrote, "and a reminder that beaver works have been altering North American landscapes for centuries."[33]

One of the finest places in which to grasp the extent of that alteration is Montana's Centennial Valley — a broad, grassy bowl, stunning in its wind-scoured austerity, ringed by a quartet of snow-frosted mountain ranges. The Centennial is filamented with streams and pocked by shallow lakes; grizzly bears trundle across its wooded shoulders and trumpeter swans splash down in its wetlands. It's the Lower 48's closest approximation of Mongolia: a place where you can gaze to the horizon without a human-built structure marring the panorama, where the US Fish and Wildlife Service warns bird-watchers to bring a full gas tank and a spare tire, where you're more likely to hear the eerie whinny of a sandhill crane than the rumble of an engine. During winter, you can practically count the valley's population on two hands.

One of the few souls brave enough to inhabit the Centennial year-round is Rebekah Levine, a high-spirited, freckled professor at the University of Montana Western. Levine — along with her husband, Kyle Cutting, a biologist who works at the valley's wildlife refuge, and their twin children — has made the Centennial her year-round home since 2010. It's an eventful place to raise a family. One winter day in 2017, a few months before my visit, the Cutting clan was playing outside when they were charged by an enraged moose cow, all flared nostrils and flailing hooves, the animal funneled toward them by a towering snowbank that ran parallel to the house. Levine, certain they'd be trampled to death, grabbed the nearest kid and dove for cover. Kyle, meanwhile, instinctively snatched up an iron rock-prying bar leaning against the side of the house and, channeling his collegiate baseball career, clobbered the advancing ungulate, who promptly keeled over in the powder. Several minutes later, with the family now watching safely through a window, the moose woke, staggered to her feet, and wandered groggily away, never to bother Rebekah and Kyle again.

That heart-stopping encounter notwithstanding, moose aren't the mammals Rebekah Levine spends the bulk of her time pondering. Levine is a fluvial geomorphologist, meaning she studies the formation of streams and rivers: how they were created, how they function, how they change over time. Inevitably, that brings her into contact with a certain aquatic mammal — "a fun little rodent that can do a lot of work," as she put it to me. And in the Centennial, where the wetlands are many and the humans few, there is no shortage of fun little rodents.

Levine and I rendezvoused in the valley on a chill late-spring day. Galloping winds swept across the prairie, driving cobalt-bodied Steller's jays to take shelter in runty willows. I was tagging along on a meeting of the Greater Yellowstone Hydrology Committee, a group of around twenty scientists who'd gathered in the Centennial for three days to talk shop. In testament to beavers' ascendancy, the creatures were a prime topic of conversation at this year's meeting — which was where Levine, the resident expert, came in.

Levine, her frizzy hair bursting from beneath a cinched safari hat, led our group along Odell Creek, a stream that twists under a cratered dirt road and spills into Lower Red Rock Lake. I noticed she wore calf-high rubber boots. "If you haven't ever walked over a beaver-impacted floodplain," she cautioned, "it feels like you're going on a wilderness expedition even though you've just jumped off the road." She kept up a patter of geological trivia as we hiked, and the visiting hydrologists hustled to stay within earshot, fighting clumps of chest-high willow that lined the eroding banks. Levine proved a gifted teacher, with a knack for interpreting landscapes for dunces like me. At one point we trekked through a shallow depression in the prairie, a faint, sinuous ditch whose outlines were stenciled in the matted grass — the ghost of streams past. "The modern channel is mostly single-thread, but the fan is riddled with paleochannels," she called over the wind. "You'll see all these abandoned oxbows and dry former wetland habitat, where there isn't any flow now but could fill up again." Some of that might have the ring of jargon, but here's the point: Beaver streams, like Odell, are messy — and that's a very good thing.

We tend to imagine rivers as blue lines meandering through valleys like sine curves wending across graph paper. I'm guilty of perpetuating this image; in this book you'll catch me referring to streams as "strings" or "ribbons." A more historically accurate simile, however, might be a meal of spaghetti — its strands writhing, intertwining, and occasionally spilling off your plate. A geomorphologist would describe such a helter-skelter river as *anabranching*; we civilians would call it chaos. Whatever your terminology, it's hard today to conceive how freely many American rivers once scampered across the landscape. In *The Control of Nature*, his riveting account of the Army Corps's grueling war against the Mississippi River, John McPhee wrote that the Big Muddy once "jumped here and there within an arc about two hundred miles wide, like a pianist playing with one hand — frequently and radically changing course, surging over the left or the right bank to go off in utterly new directions."[34]

If a dynamic river possesses the hands of a pianist, then beavers had turned Odell Creek into a virtuoso. Although the stream currently flowed through but one visible course, Levine told us the floodplain was scarified with a maze of historic channels, between which Odell had hopped over the millennia like an impatient driver changing lanes. Beavers were largely responsible for that dynamism: By slowing down flows, capturing sediment, and raising the height of the stream's surface, their dams had repeatedly forced Odell's waters onto its surrounding floodplain, furnishing meadows with nourishing sediments and transforming the stream into a bewildering maze of wetlands, side channels, and meanders — from single thread to proverbial spaghetti bowl. Even the brushy, pitiful vestiges of neglected dams changed the stream's course, scouring holes and deflecting flows. One Colorado study found that islands often form behind abandoned dams, forcing the channel to split and turning single-thread streams into braided ones.[35]

"One of the coolest things that beaver dams do is fail," Levine said with an admiring nod to one half-blown dam, its constituent sticks shuddering slightly in the current. "These structures are actually affecting sediment transport and channel movement and creating dynamism in these systems."

Beaver country, with Levine as guide, was rife with signs and patterns, the sort of telling details that Annie Dillard called "unwrapped gifts and free surprises" — an eroding bank here, the ghost of an old dam there. It helps, of course, to be a geomorphologist, but it matters more that you *see*, that you attend to the clues that together tell a river's story. Levine paused at a *point bar* — a sandy beach, deposited on the inside of a riverbend, on which you might land your canoe for a picnic. She called our attention to the peeled willow stems that littered the bar.

"Basically all of the willow sticks on point bars in this creek, guess where they came from? Beavers," Levine said. "They're extremely messy builders, and there's a constant flux of sticks downriver when they're working.

Willows can sprout from these cut stems, and that's adding even more complexity to the habitat." As beaver-transported willows took root on the point bar, they forced the creek toward its outer bend, where it chewed into the bank and carried off still more earth, which would later settle behind downstream dams. The saga's details change in every stream, but the overarching reality is this: Beavers are agents of profound change, responsible for sculpting streams' forms and dictating their functions.

Although it seems unmistakable today, the notion that beavers shape entire watersheds was not always obvious. Among the pioneers of fluvial geomorphology was a Berkeley researcher named Luna Leopold, son of the legendary ecologist Aldo Leopold. Leopold was a visionary whose fastidious measurements did much to help us understand how rivers work, yet he had a blind spot. In his seminal paper *River Flood Plains: Some Observations on Their Formation*, beavers don't earn a mention.[36] The broad, grassy floodplains that flank many rivers, wrote Leopold and his coauthor, the geographer M. Gordon Wolman, owe their existence to "material deposited from high water flowing or standing outside of the channel" — overbank floods produced by rainfall or snowmelt. That streams often rose so high thanks to the activities of a knee-high rodent escaped their attention. "The classic story of how valleys fill is a meandering stream going back and forth, depositing sediment," Chris Jordan, a beaver-admiring fisheries biologist with the National Oceanic and Atmospheric Administration, told me. "But with beavers around, the historical landscape processes would have been completely different."

It's hard to begrudge Luna Leopold his oversight. The stream flow data and field observations he drew upon were collected after 1900, long after beavers had disappeared from nearly all of America's lands and waters.[37] As the animals rebound from the fur trade, we've had to update our conceptual models to account for their influence. In 1980, for instance, the field of aquatic ecology came to be dominated by "the river continuum," the notion that waterways transition along their course, seamlessly and predictably, from steep, forested headwaters to open valley bottoms. Three decades later, however, an engineer named Denise Burchsted proffered a different model: the river *discontinuum*, which held that pre-colonization streams were disrupted along their length by glacially scoured holes, downed trees, and, most of all, beaver dams. Rather than free-flowing chutes, Burchsted wrote, historical creeks were patchy networks of ponds, meadows, and braided channels — only fitfully connected upstream and down, but inseparable from the floodplains that bracketed their banks.[38] Many natural river systems seemed to blur the line between land and water: Floodplains were less discrete landscape features than ecotones, fuzzy transitional worlds where wet bled into dry and turned everything wondrously damp.

Although North America may never again host its full beaver complement, we can still envisage the soggy world that once prevailed. In 2005 David Butler and George Malanson, geographers at Texas State University and the University of Iowa, respectively, calculated that somewhere between 15 and 250 million beaver ponds puddled North America before European arrival.[39] Given the continent's diverse topography, there's really no such thing as a "typical" beaver pond: Researchers in Montana's Glacier National Park surveyed ponds whose area averaged a skimpy tenth of an acre, while others in eastern North Carolina found that average ponds measured a robust four and a half acres. For argument's sake, though, let's split the difference and estimate that the continent was laced with 150 million ponds that averaged a single acre apiece. If that's true, beavers once submerged 234,000 square miles of North America — an area larger than Nevada and Arizona put together.

Thanks in part to beavers, ours was a watery country, a matrix of ponds and swamps, marshes and wetlands, damp mountain meadows and tangled bottomlands. But the same luscious fur that made beavers so well adapted for this aquatic world would soon prove their downfall — and the undoing of the ecosystems they helped create.

— CHAPTER TWO —

Dislodged

I f you were a fur trapper roaming the little-mapped American West during the 1830s, the Rocky Mountain Rendezvous would have been the
most thrilling — and probably the only — event on your social calendar. The
Rendezvous was part bazaar, part reunion, part rodeo. Each summer hirsute
trappers trickled down from the mountains like meltwater to valley encamp-
ments in Wyoming, Idaho, and Utah, where they swapped a year's worth of
"hairy banknotes" for gunpowder, tobacco, and other essentials. The confabs
usually devolved into pandemonium; many attendees gambled and drank
away their annual earnings in days. John Kirk Townsend, an ornithologist
who tagged along one year, bellyached about "the hiccoughing jargon of
drunken traders, the *sacré* and *foutre* of Frenchmen run wild, and the swear-
ing and screaming of our own men . . . being heated by the detestable liquor
which circulates freely among them."[1] Guns and booze made for a volatile
combination. "We who were not happy" — i.e., plastered — "had to lie flat
upon the ground to avoid the bullets which were careening through camp."[2]

Rocky Mountain Rendezvous still convene today, although they're
considerably tamer affairs: When I attended the Green River gathering
in Pinedale, Wyoming, I didn't have to dodge a single projectile. Pinedale
is the perfect place to take in a modern Rendezvous, which is essentially
a Renaissance fair for people who prefer buckskins to tunics. The town,
which abuts the forbidding Wind River Range, has turned fur trade nos-
talgia into a cottage industry. Pinedale's main attraction is the Museum of
the Mountain Man, and the annual Rendezvous probably doubles its two-
thousand-person population. Even its slogan yearns for an era when the
maps were murkier, the mountains wilder, and the human lives cheaper.
Welcome to Pinedale: All the civilization you need.

One crisp morning, I strolled the fair's green, looking to talk beaver. Although most patrons of modern Rendezvous seem to be midwesterners RVing to Yellowstone, some of its vendors are legit trappers hawking furs of their own procurement. Rows of canvas tents surrounded the lawn, pelts dangling in their dark interiors. I ducked into one well-stocked tent and walked the far wall, running my fingers over merchandise that hung from rings passed through lifeless snouts: gray fox, red fox, coyote, bobcat, striped skunk, otter with feet, otter without feet, raccoon, three different sizes of ermine. A stack of beavers, coats shading from balsa to cocoa, lay piled like flapjacks on a side table.

The proprietor, an elfin man with sandy hair and a Clausian beard, bounded up to greet me. His name, he told me, was Don Cooper, though he sometimes went by Missing Link. "I've also been called Hairball, Furball, Crack, Spider Bite, and Missing Britches," he drawled. Unfortunately I couldn't reciprocate with my own colorful nicknames, but he promised to answer my beaver questions nonetheless.

Missing Link's tent overflowed with kitschy gewgaws: beaded purses, arrowheads on thongs, wooden pistols in leather holsters. His beaver knowledge, though, was authentic and unassailable. Cooper spends much of his year on the road, buying and selling pelts on a national Rendezvous circuit that loops from Montana to Wyoming to Oklahoma. But he still finds time to trap nuisance beavers at home in Texas, particularly in February and March, when late-winter rains raise pond levels and make landowners leery. In summer, he added, folks appreciated beavers because they kept ponds full, which made me question the foresight of Cooper's clients. He enticed beavers with a smelly lure of his own concoction, a blend of castoreum, glycerine, cinnamon, and some ingredients that remained proprietary. When a resident rodent, motivated by his innate territoriality, swam over to investigate the scent of the strange trespasser on his turf, he'd find himself fatally clenched in one of Cooper's steel body-gripping traps or entangled in a snare. Cooper's fellow trappers have recognized the potency of castor-based lures for at least three hundred years. "It is this deposit," wrote Osborne Russell of castoreum, "which causes the destruction of the beavers by the hunters."[3]

Like most contemporary trappers, Cooper earned far more money exorcising beavers from backyards than he did selling their pelts. Eight days of nuisance beaver trapping on your ranch, he told me, would run you around six hundred dollars. But he seemed to trap as much for pleasure as for profit. "I had one pond that belonged to a friend of mine that I knew we had a beaver in there, and I knew it had one toe missing because I'd caught it by the toe and it pulled the toe off," Cooper told me as customers filed in and out. "Three years later, I went back and caught the beaver that was missing that toe. I had that pelt tanned and I gave it to the landowner. He was the best man when I'd got married. He was really pleased. Him being so pleased made me feel good."

We walked over to his pile of pelage. Cooper nodded warmly at a couple browsing the necklace section. "How do, folks, good to see you," he said. He stooped to leaf through the mound of mammals. Some of the pelts came from beavers he'd caught and dressed himself; others he'd purchased from colleagues. After he caught a beaver in a snare or a trap, he skinned it, fleshed it, and stretched the pelt out taut to dry, like the surface of a trampoline, on a wire hoop. Not all the beavers he trapped went through the wringer. Some he left in the woods for the coyotes.

"If the diet's been bad, or they've been sick, and they got poor fur quality, it's just not worth the time to skin 'em," he said as I caressed the heap of furs. The thick skins, which once kept the cold and damp off living rodents as they nibbled willow in Arkansas and Michigan and Utah, were impossibly plush beneath my fingers, the texture of warmth and wealth. A little spasm of covetousness passed through me. "You spend thirty minutes for four dollars or thirty minutes for eight dollars, and it don't make sense on the four dollars."

Cooper flipped a pelt, exposing its tanned underside. He peered at the skin through wire-rimmed specs, interpreting its owner's biography as though reading a palm. "That white mark in there, this is scars from where the beaver had been fighting with other beavers," he said. "And this one here is another Utah beaver. You can see the length of the fur — they get real soft guard hairs in Utah." He turned the skins in his hands. He knew who'd dressed each fur by the quality of the tiny holes along the skins' fringes, where the beavers were stretched on their hoops. "These here punches are a little too far apart," he opined with a frown.

The irony of the latter-day Rendezvous is that there are thousands more people reliving the mountain man era today than ever existed actual mountain men. The Rocky Mountain fur trade cycled through boom and bust nearly as fast as a dot-com bubble: Just fifteen years after the first Rendezvous, in 1825, the gatherings collapsed for want of beaver. Most of the men we associate with the fur trade operated during this era: Jim Bridger, Kit Carson, Jedediah Smith, and Hugh Glass, whose mauling by a grizzly was graphically immortalized in *The Revenant*. "There is, perhaps, no class of men on the face of the earth, who lead a life of more continued exertion, peril, and excitement, or are more enamored of their occupation, than the free trappers of the West," wrote Washington Irving.[4]

In reality, though, the mountain men were late entrants to the beaver business. Since the 1500s, trappers and traders had been harvesting North America's beavers for coats, capes, and, most of all, hats. Bridger and his comrades were the final executors of a three-century killing spree. The drive for pelts paved the way for towns and farms to plow under the very wilderness in which the mountain men made their bones. The industry also decimated indigenous communities, acting as the vanguard — sometimes unwittingly, sometimes deliberately — for remorseless politicians, military leaders, and businessmen who yearned to seize the West for whites. "These adventurers acted the part of precursors as well as trappers, and went in advance of civilization, and discovered countries now occupied by the agriculturalists and mechanics," wrote David Coyner in 1847.[5] The fur trade, like so many extractive industries, devoured itself — and reshaped North America in the process. Watching Don Cooper fondle his pelts on the Pinedale green, I recalled Oscar Wilde: *For each man kills the thing he loves.*[6]

"Canada," Margaret Atwood wrote, "was built on dead beavers."[7] The tireless rodent certainly looms large in the psyches of our northern neighbors — during the closing ceremonies of the 2010 Vancouver Olympics, inflatable beavers with sumo wrestler physiques blimped out onto the ice. Along with the maple tree, beavers are an official emblem of Canada; the association is so strong that it's cemented in the creature's specific name, *canadensis*.

Although Americans don't pay similar heed to our beaver heritage, Atwood could have penned the same line about the United States. Beaver furs were the wind in the sails of the *Mayflower*, furnishing Pilgrims with a tradable good with which to repay their creditors back in England. "The Bible and the beaver were the two mainstays of the young colony," James Truslow Adams wrote. "The former saved its morale, and the latter paid its bills, and the rodent's share was a large one."[8]

More than timber, cod, or any other natural resource, beavers help explain just about every significant American geopolitical event between European arrival and the Civil War. The American Revolution? Incited, in part, by the much-maligned Proclamation of 1763, a royal fiat that barred colonists from settling west of the Appalachians in order to prevent them from disrupting the British trade in furs.[9] The War of 1812? Egged on by Canadian and American traders sparring over control of beaver-rich lands around the Great Lakes. The wars that drove France to surrender its New World colonies? Inflamed by conflict in the Ohio Valley, a vital link in the continent-spanning chain of French fur outposts. Decades before James Polk whipped up Manifest Destiny to justify his imperialist inclinations, beavers were impelling colonists westward: The desire to claim new trapping grounds motivated Thomas Jefferson to purchase the Louisiana Territory and dispatch Lewis and Clark to explore it.

Needless to say, white settlers were millennia short of being the first North Americans to depend on beaver. For thousands of years, the continent's native peoples had eaten the rodents' flesh and fatty tails, draped themselves in furs, and used castoreum as medicine; in 1609, for instance, the French author Marc Lescarbot observed members of Nova Scotia's Mi'kmaq placing slices of "beaver stone" atop wounds.[10] The Naskapi burned beaver scapulae and searched cracks in the scorched bones for portents; British Columbia's Clayoquot employed beaver molars as dice in gambling games; and numerous tribes and First Nations used beaver mandibles and incisors as chisels, scrapers, and knives. One Cree creation story centers around Great Beaver, whose titanic dam flooded the entire world, compelling Wisagatcak the trickster to bundle other animals onto a raft and replant the planet with moss.[11]

The arrival of Europeans warped indigenous peoples' relationship with beaver from subsistence and kinship to extraction, turning many tribes, as Harold Hickerson put it, into "a vast forest proletariat whose production was raw fur."[12] The trade began innocuously enough in the 1500s, with European cod fishermen casually exchanging goods for pelts during their spells ashore in Newfoundland. When Henry Hudson found the Mohawk eager to trade in 1609 in modern-day New York, the rush was on. Europe was then in the process of exterminating nearly all its own beavers, and the untapped source of pelts proved irresistible. Soon the English, Dutch, French, and even the Swedish were competing to control eastern arteries like the Hudson, Delaware, St. Lawrence, and Kennebec, funneling furs downriver from New England's interior to the sea. When the Dutch bought Manhattan from the Lenape in 1626, the island itself was little more than a pot-sweetener: The real prizes were the 7,246 beaver skins that sailed to Europe aboard the *Arms of Amsterdam*.[13]

Beavers have the unfortunate habit of betraying their own presence: The very dams and lodges that protected the rodents for millions of years became unmistakable billboards, advertising active colonies to trappers. "The most simple method," wrote one traveler, "is by destroying these houses, and draining the ponds on which they are situated, so that the animals, being alarmed and deprived of the water so necessary to their existence, take immediately to their flight and become an easy prey," slain with clubs, arrows, or nets.[14] In the early 1700s the English introduced steel traps and castoreum lures to the trade, a foolproof combination that by century's end had found their way into the kit of most every trapper and led, wrote the historian Harold Innis, to "the exhaustion of the beaver fields."[15]

Since most Europeans lacked the skill and patience to catch beavers themselves, they bought pelts from Indians in exchange for knives, hatchets, kettles, cloth, beads, and — despite sporadic prohibitions against distributing these forms of currency — liquor and guns. Although many Indians were shrewd negotiators who played traders against each other to reap better deals, unscrupulous whites often enmeshed their native partners in webs of usury. Many indigenous people sold their land for pennies on the dollar to escape trading debts. Countless others lost their lives to war and disease. One especially ghastly smallpox epidemic, which originated on a steamboat full of Missouri River fur traders in 1837, tore through half a dozen tribes, including the Sioux, Blackfeet, and Mandan. By the time the virus ran its course, some seventeen thousand Indians lay dead, and their societies in ruins. "The prairie had become a graveyard," wrote one observer, "its wild flowers bloom over the sepulchers of Indians."[16]

The frivolous accoutrement that spawned this carnage was the *beaver hat* — a phrase that suggests, perhaps, a bulky head-warmer, akin to a Russian ushanka or Davy Crockett's coonskin cap. In fact, beaver hats tended to be elegant felt creations, with names like the Regent, the Wellington, and the Paris Beau, that crowned well-heeled aristocrats. Nearly any fur, rolled and compressed, can be transformed into felt — but it's the barbed hairs of the beaver's underfur, which interlock like Velcro, that make the most durable, waterproof, and malleable head-toppers. "Many men died, a continent was explored, an indigenous race degraded and its culture crushed," wrote the historian Don Berry, "all because beaver fur, with its tiny barbs, felted up better than any other."[17]

Motivated by Europe's sartorial cravings, the fur trade pushed west, driven by strapping, good-humored French Canadian voyageurs piloting slender birch-bark canoes in quest of *le castor*. (The French generally treated their native colleagues more squarely than the English, though Britain set a low bar.) Throughout the seventeenth and eighteenth centuries, the industry was a frenzied free-for-all characterized by shifting alliances, geopolitical intrigue, and brutal slaughter of both beavers and humans. The Iroquois massacred the Hurons, the Dutch clashed with the Swedes, the Chippewa routed the Miami, and the British constantly butted heads with their archnemeses, the French. The only constant remained the prodigious supply of pelts. In 1700 the market grew so glutted that warehouse owners in Montreal burned three-quarters of their inventory, spreading "an almost unbelievable stench and gloom of dark greasy smoke over the town."[18]

Nearly every North American stream, river, and lake seemed to brim with beavers — at least, until the marauding traders rolled through. Around Lake Erie, trappers routinely captured thirty in a single night. In the Delaware Valley an English traveler remarked that the entire country "aboundeth with beavers, otters, and other mean furrs."[19] John Bakeless, in his book *The Eyes of Discovery*, wrote that "beaver filled the inland rivers, creeks, and ponds all the way to the Rockies."[20] David Thompson, a surveyor who arrived in Hudson Bay in 1784, avowed that North America above the 40th parallel "may be said to have been in the possession of two distinct races of Beings, Man and the Beaver."[21] And in New York's Adirondacks, wrote Harry Radford, "it is evident that every lake and pond was occupied, and every river, brook, and rill, from the largest to the most insignificant, thickly peopled with these industrious and prolific animals. They seem to have completely possessed the land, and to have been abundant almost beyond our present conception."[22]

These white explorers found a continent that not only teemed with beavers, but had been profoundly shaped by them. Many river valleys were so clogged with dams and ponds that they defied navigation, forcing parties to stick to broad, deep watercourses in canoes or traverse the uplands on foot. The Lehigh River, a Delaware tributary, was "almost choked" with beaver dams, which, according to Bakeless, helped form the "Great Swamp" at its headwaters, a dismal stretch that went by the name "Shades of Death."[23] Louis Hennepin, a Belgian-born priest who explored the Midwest in 1679, remarked that the beaver-filled Kankakee River was characterized by "marshy Lands, which are so many quagmires, that one can scarcely walk over them." Had the ground not been frozen, Hennepin's party wouldn't have been able to find a dry campground.[24]

A similar tale comes from Pierre Esprit Radisson, a Canadian who was captured by the Mohawk at the age of sixteen, escaped to become a trader, and eventually helped found the Hudson's Bay Company, a British government-sanctioned fur monopoly that persists today as a chain of retail stores. On an expedition near Lake Superior in 1661, Radisson and his companions came upon a beaver complex so vast that it awed even the experienced woodsman. "What a wonderfull thing to see the industrie of that animal, [which] had drowned more than 20 leagues in the grounds, and cutt all the trees," Radisson wrote. Although his party knocked down dams to clear passage for their canoes, the terrain turned treacherous. Silted-up beaver ponds had grown carpeted with "moss [which] is two feet thick or there abouts," forming a "trembling ground."[25] "If you take not care you sinke down to [your] head or the midle of [your] body. When you are out of one hole you find yourselfe in other," the Canadian added. Radisson might not have made it without a tip from his native companions, who advised him to disperse his weight across the ground's surface and crawl along "like a frogge."

Although Lewis and Clark never resorted to Radisson's awkward method of locomotion, they, too, were astonished by beavers' works. In midsummer 1805 the Corps of Discovery, the duo's exploration party, followed the Missouri River through present-day Montana, passing sheer bluffs and cottonwood bottomlands teeming with elk, sheep, and bison. "Wherever we find timber, there is also beaver; Drewyer killed two today," Lewis wrote on June 30.[26] The rodents later exacted revenge: One wounded beaver bit Clark's dog, Seaman, in the hind leg, almost fatally. Another gnawed down the pole on which Lewis had affixed a note for his comrades at a river fork, leading Clark astray for a mile, much to his irritation.[27]

The farther up the Missouri the corps pushed, the more impressive the beaver's influence became. On July 18, wrote Lewis, Clark explored a tributary and "saw a number of beaver dams succeeding each other in close order and extending as far up those streams as he could discover them in their (course) towards the mountains."[28] A week later, as the group negotiated a tricky stretch of narrow channels and brushy archipelagoes, Lewis, a keen-eyed naturalist, wrote that beavers "compell the river in these parts to make other channels . . . thus the river in many places among the clusters of islands is constantly changing the direction of such sluices." All in all, beavers appeared to be "very instrumental in adding to the number of islands with which we find the river crouded."[29]

On July 30, while ascending the Jefferson River, Lewis encountered such a morass of beaver dams and ponds that he was forced to wade "up to [his] waist in mud and water" to escape the valley.[30] Even that trying experience did not diminish his admiration. Three days later he reported seeing beaver dams that stood five feet high and flooded several acres apiece. "These dams are formed of willow brush mud and gravel," Lewis marveled, "and are so closely interwoven that they resist the water perfectly." Trying to emulate such structures, he added, "would puzzle the engenuity of man."[31] Nearly every day that month saw "the beaver pleanty as usal," and as often as not the corps's diaries contain references to dams, ponds, felled trees, and the large-scale manipulation of streams and rivers.

No sooner did Lewis and Clark return to St. Louis, the frontier's largest hub, than the loose-lipped explorers began spreading mouthwatering tales of the Missouri's bounty. Enticed by "campfire bull sessions, barroom yarns, [and] refined after-dinner conversation over cognac and cigars,"[32] waves of would-be trappers rushed upriver. Back east, the business of hunting beavers had been carried out by Indians, allowing white traders to reap profits without dirtying their hands. As the industry's theater shifted west, however, the dynamics changed. Tribes like the Blackfeet and Crow, whose cultures and cuisines revolved around bison, had little interest in pursuing rodents. As one trader put it, the Plains tribes "considered the operation of searching for [beavers] in the bowels of the earth, to satisfy the avarice of the whites, not only troublesome, but very degrading."[33]

Annoyed traders predictably chalked up the tribes' reluctance to indolence. But the real explanation was a religious one — with a profound connection to beaver ecology. Today few people understand that link better than Rosalyn LaPier, an ethnobotanist and Blackfeet member raised on the tribe's reservation in northwest Montana. LaPier grew up steeped in Blackfeet tradition; her grandfather was a ceremonial singer fond of paying his granddaughter pennies to recite her family's genealogy. From her grandparents, she learned early that the Blackfeet believe in three separate realms of existence — the sky, the earth, and the water world, the latter inhabited by Soyiitapi, or water beings. "I was raised to be really respectful of the water world and the supernatural entities that lived there," LaPier told me. And no water being commanded more respect than Kitiaksísskstaki, the Not Real Beaver.

Although she studied physics in college and worked for a time in tribal energy development, LaPier eventually found her way back to Blackfeet religion, completing her PhD in history in 2015 at the University of Montana. Her dissertation is a complex explication of how the Blackfeet regarded nature — not as a capricious force that acted upon a helpless people, but as a responsive partner whose fluctuations could be controlled. Beavers were perhaps the most valuable partners of all. In one tribal story, writes LaPier, the Not Real Beaver invites a human to winter with him in his lodge, where the supernatural being schools the person in nature's ways, introduces him to "natural elements" like tobacco, and grants him "power of the waters." "The following spring," LaPier adds, "the beaver 'transferred' this supernatural knowledge and material objects to the human, who in turn shared them with other humans."[34] The beaver also bestowed a medicine bundle — a collection of sacred objects, such as tobacco seeds and animal skins — and granted humans mastery over the bison, the tribe's most important food source. To LaPier's mind, the beaver's centrality within stories and ceremonies encodes a powerful ecological message: "Beaver are important, extremely important, to living in an arid environment," she told me. They created the watering holes to which game flocked and the oases in which plants flourished. They even cut and peeled the driftwood women gathered to fuel fires.

The Blackfeet's respect for beavers manifested in strict prohibitions against killing: In one story a man who attempts to slaughter a beaver has his wife stolen away by the angry creature.[35] No wonder, then, that the Blackfeet declined to participate in the beaver trade. As the Canadian anthropologist R. Grace Morgan put in 1991, Plains tribes avoided hunting beaver in "response to the limited availability of surface water and a recognition of the beaver's role in maintaining these resources."[36] In the northern woodlands, Morgan added, tribes like the Cree operated under no such cultural restrictions — not only because they valued the rodents' pelts and meat, but because flooding in the forest was an obstacle to hunting, not a blessing.

"From here in the twenty-first century, we can look back and see that people then really did have a much more sophisticated understanding of their ecology than we recognize today," LaPier said. Western scientists may have taken centuries to come around to the notion that beavers create life, but to the Blackfeet and other tribes the animal's vitality was always self-evident.

Unable to rely on their "forest proletariat," thousands of white men took to the hills to prosecute the hunt themselves — and swiftly picked the West clean. In 1843 John James Audubon traveled twenty-two hundred miles along the Missouri in search of mammals to paint. Though he saw tracks, heard a tail-slap, and dismantled a lodge, he didn't see a single beaver. Even the mountain man in his employ couldn't procure him a specimen. The land, the artist wrote, was "quite destitute."[37]

On a continental scale, the fur industry was far from moribund: The Hudson's Bay Company had its biggest pelt-trading year in 1875, vending more than 270,000 furs, nearly all from Canada. But the beaver's days as one of the preeminent drivers of American culture, economy, and politics were over. "The one true thing about every American frontier that seems concrete and immutable," the essayist Charles Pierce has written, "is that it does not last."[38]

―――――

Truthfully, Audubon shouldn't have been surprised to find the Missouri barren. For more than two centuries, a conquering army of white trappers and traders had ravaged beavers in every stream and pond they encountered, leaving ruins in their wake. Beaver numbers in New England began tapering off as early as the mid-1600s; by the eighteenth century, Massachusetts was beaver-free. Connecticut, Vermont, and several other northeastern states soon joined the ranks of the beaverless. Among those who bemoaned the destruction was Thoreau; the hermit, who considered beavers one of the "nobler" animals, wrote that the rodent's destruction, along with the eradication of cougars, bears, moose, and other species, had "emasculated" the Northeast.[39]

If the first wave of trappers encountered a continent molded by beavers, the farmers and settlers who followed met a landscape shaped by their absence. "The elimination of the beaver . . . left New England with a wealth of place names that no longer made much sense: scattered across the map of the region one still finds Beaver Brooks, Beaver Stations, Beaver Creeks, and Beaver Ponds," wrote William Cronon in *Changes in the Land*.[40] And that's just English: Ahmeek, Michigan; Capa, South Dakota; and Kinta, Oklahoma, all derive their names from native words for beaver.

For all the nomenclatural upheaval, the ecological repercussions were far more severe. Many beaver-formed wetlands are inherently liminal, transitioning ceaselessly from one state to the next; a beaver complex is one of the few opportunities we mortals have to watch geological processes unfold within the duration of a single human life span. Often the progression runs from water to earth, as beaver ponds slow flows, trap sediment, and gradually fill with silt and pioneering plants, whereupon their creators move upstream to begin the cycle again. Left behind are open, grass-filled meadows — their surfaces flat, treeless, and flan-like underfoot, bermed with the overgrown contours of long-ago dams. The fecundity of these sunlit havens wasn't lost on hunters. "Beaver meadows are splendid feeding grounds for deer and other animals," remarked James Campbell Lewis, a hunter and trapper who went by the name Black Beaver. "I have seen beaver meadows . . . covering hundreds of acres."[41]

When trappers de-beavered North America, they threw that natural evolution into fast forward. As dams deteriorated and ponds drained, they left behind some of the finest soil a farmer could till, a newly exposed mélange of "leaves, bark, rotten wood and other manure," as Jeremy Belknap put it in his *History of New Hampshire*. "The whole tract, which before was the bottom of a pond, is covered with wild grass, which grows as high as a man's shoulders, and very thick. . . . Without these natural meadows, many settlements could not possibly have been made."[42] An English traveler named Henry Wansey corroborated the account. "It is a fortunate circumstance to have purchased lands where these industrious animals have made a settlement," Wansey wrote. "At some of them, there have been four ton of hay cut at an acre."[43] When beavers were trapped out of the American Southwest more than a century later, the abrupt dam failures released so much water that the Gila River Pima Indians recorded the first major flood in their oral history.[44]

To understand how dramatically landscapes unraveled in beavers' absence, consider the research of Johan Varekamp. Varekamp, a lean, fast-talking Netherlands native, is a geochemist at Wesleyan University in Middletown, Connecticut; his wife, Ellen Thomas, is a micropaleontologist at nearby Yale. In the 1990s the duo organized a series of research cruises in Long Island Sound, the wedge of ocean that squeezes between Connecticut's shoreline and the tapering moraine of Long Island. Using a bottom-penetrating coring tube, Varekamp and Thomas hauled up cylindrical samples of muck, which they analyzed for everything from pollen to salinity. But it was the mercury concentrations, which spiked near the mouth of the Housatonic River, that grabbed their attention.

The source, they learned, was the city of Danbury, a former capital of fur hat manufacturing. Hatmaking once entailed an act of appalling chemistry: Hatters brushed beaver or rabbit pelts with mercury nitrate, an orange solution that matted fur into malleable felt. Many hatters who handled the potent neurotoxin suffered from mental fogginess and uncontrollable twitching — a disorder, known as the Danbury Shakes, that likely spawned the idiom *mad as a hatter*.[45] Not until 1941, after years of worker advocacy, did the state finally ban mercury's use in hatting. By then the damage to human health and the environment had been done: On the grounds of one former hat factory, Varekamp found that mercury concentrations remained triple the state's standard.[46]

But Varekamp's rodent-related investigations had only begun — for it wasn't just the fur trade that left behind its signature in the Long Island Sound's sediment, but New England's beavers themselves. When he and Thomas examined their layers of muck, they found that diatoms, a kind of silica-based algae, had exploded around 1800. That made sense: Expanding European colonies and their livestock would have fouled the sound's waterways with runoff around that time, fertilizing the ocean with nitrogen-rich waste and birthing algal blooms. To their surprise, however, they also found evidence of an earlier spike in algae, in the late 1600s, that appeared to be unrelated to human population. Varekamp scoured history for an explanation. He remembered mercury; he thought about hats and pelts and, finally, beavers.

The story, he told me, goes something like this. Southern New England's earliest European explorers — including Adriaen Block, Varekamp's fellow Dutchman, who sailed to the region four times between 1611 and 1614 — instantly realized its fur potential. By the mid-1600s around eighty thousand pelts were leaving the region each year, bound for Holland and England.[47] As beavers vanished, the same epic collapse of unmaintained dams that created peerless farmland also conveyed massive quantities of pond sludge — decayed leaves, sticks, insects, and other detritus — to the sea. The result was a surge of fertilizer that Varekamp and Thomas termed the "beaver peak."[48] "You have all these beaver ponds that are ultimately going to hell," Varekamp told me, "and all that decaying wood and organic matter ultimately leads to a flux of nutrients to the sound and a pulse of algal blooms."

In other words: Beavers were so omnipresent, so influential, in the Northeast that their destruction can still be detected in the sea itself.

Colonists quickly choked New England with dams of their own, clogging Connecticut's rivers alone with more than four thousand walls to power saw-, grist-, and paper mills. Even so, Varekamp said, Long Island Sound's watershed still delivers several times more sediment to the sea than it did before white settlement. Partly that's because deforestation destabilized soils and increased erosion; but partly it's because, for all our dam building, we can't hope to replicate the sediment-storing feats of our flat-tailed forebears. "Every secondary and tertiary river would have been totally dammed," Varekamp said with a touch of melancholy. "It is hard to imagine what the overall landscape would have looked like — how gorgeous it was, how green, how lush."

Among the fur trade's many injustices is how little we know about the fully beavered North American landscapes that prevailed before trapping. Although many mountain men were erudite, most were not devoted journalers, and, for all their woods-sense, they certainly weren't trained scientists. While some trapping expeditions carried naturalists, the hapless researchers were usually considered a drag on profits. In one letter Peter Skene Ogden of the Hudson's Bay Company griped about having to take "2 in quest of flowers 2 killing all the birds in the Columbia & 1 in quest of rocks and stones . . . they are a perfect nuisance."[49] Thomas Nuttall, an English botanist who journeyed up the Missouri in 1811, earned the nickname *le fou* — the fool — from boatmen for his daft affinity for plants.[50] "What valuable and highly interesting accessions to science might not be made by a party, composed exclusively of naturalists," lamented John Kirk Townsend, after he was pulled away from a promising oasis before he could properly assay it, "on a journey through this rich and unexplored region!"[51]

Lacking keen observations from early naturalists, it fell to twentieth-century scientists to retroactively deduce how beavers molded the continent. The most astute was Rudolf Ruedemann, the state paleontologist of New York, who, in the 1930s, grew fascinated by the "perfectly leveled" land east of Troy. The valley, nine miles long and half a mile wide, seemed too broad to have been carved by the modest creeks that meandered across its bottom. A glacial origin didn't square with the evidence, either. The paleontologist believed that a "third agent" must have created the gently sloping plain: beavers.

The animals, Ruedemann postulated, had begun constructing their dams at the valley's foot; as each successive pond filled in, they marched a bit farther upstream to build afresh, repeating the process for as many as twenty-five thousand years until the valley was blanketed in a deep tract of smooth, fertile soil. When beavers were trapped out from this and other valleys, they left behind a subtle yet monumental legacy. "The fine silt gathered in the beaver pools has produced the rich farm land in the valleys of the wooded areas of the northern half of North America," Ruedemann wrote.[52] Seventy years later Colorado State University researchers Lina Polvi and Ellen Wohl corroborated Ruedemann's "beaver valley hypothesis" on the other side of the country, finding that up to half the sediment in one Rocky Mountain meadow comprised fine-grained particles that could only have settled in ancient beaver ponds.[53]

Ruedemann aside, most scientists couldn't wrap their heads around the grandeur of castorid influence. Ecology, the study of how organisms interact with one another and their surroundings, is a young discipline; not until 1935 did the word *ecosystem* enter the parlance, a century after North America's beavers had been virtually obliterated. Although researchers recognized that beavers engineered individual ponds, the possibility that they'd shaped a continent — that the geological mass we call North America might, as Frances Backhouse put it, more accurately be termed Beaverland[54] — was hard to grasp and harder still to quantify.

Ultimately, it would take a researcher at an oceanographic institution to untangle the mysteries of North America's interior.

Bob Naiman's beaver obsession began in the backwoods of Quebec, during a futile search for the perfect stream. In the late 1970s Naiman, a biologist, was working for the Woods Hole Oceanographic Institution, trying to figure out how habitat and food affected Atlantic salmon production. To accomplish that, he needed to find "reference reaches" — pristine stretches of stream, untouched by humans, that he could compare with other waterways. Naiman traipsed the Canadian woods in a futile hunt for references. Everywhere he traveled, though, had been altered beyond use — not by people, but by beavers, which had faced only light trapping in the remote region. Finally it dawned on Naiman that beavers weren't a confounding factor in the story of Quebec's watersheds — they *were* the story.

"It was so obvious, and it had been in front of me for so long," Naiman recalled to me. "And the implications just kept getting better and better."

There was only one snag: Naiman's new vision — to explore how beavers had altered the geology, ecology, and hydrology of a ten-thousand-square-mile watershed — seemed impossibly ambitious. "Literally everybody in the freshwater community said it couldn't be done," Naiman told me. But Naiman worked, after all, for an oceanographic institution, where many colleagues were examining vast and abstruse processes like the circulation of the Indian Ocean. "Ten thousand square miles was nothing compared to what these guys were doing."

Over the next several years, Naiman executed what remains one of the most colossal studies of beaver influence ever attempted. He and his team dug up sediment cores and measured the debris embedded in streams; scooped water samples and trapped insects; stuffed wood in mesh bags to watch it decompose and captured methane bubbles in bell jars. The results suggested that beavers had sunk their imposing incisors into virtually every aspect of the Quebecois landscape. Small streams were interrupted by a whopping 10.6 dams per kilometer — a dam every football field. (Prolific though that sounds, researchers in Wyoming have found streams jammed with up to *fifty-two dams* per kilometer, or one every twenty yards.[55]) Individual ponds captured enough water to inundate more than a city block. By drowning and cutting down trees, beavers permitted sunlight to penetrate dense forest canopies, stimulating photosynthesis and growing crops of the microscopic plants that buttress aquatic foodwebs. Some ponds trapped sixty-five hundred cubic meters of sediment, enough to fill more than two Olympic swimming pools. Altogether, Naiman calculated, the region's beavers were storing 3.2 million cubic meters of sediment — enough to gag every stream in the watershed with more than a foot of sludge had the rodents not been around to capture it.[56]

Spurred by his Quebec studies, Naiman moved on to even grander beaver research. The biologist got his hands on sixty-three years of aerial photographs that surveyors snapped of Minnesota's Kabetogama Peninsula. When the photo series began in 1927, just sixty-four dams dotted the peninsula. By 1986, after stringent trapping regulations, the disappearance of wolves, and the regrowth of aspen had helped beavers recover, Naiman could count 835 dams — more than a tenfold increase in a few decades. Initially, less than 1 percent of the landscape had been impounded by beaver ponds; now 13 percent of the peninsula was underwater.[57] Vegetation communities had evolved accordingly, from bog to marsh to forested wetland and back again, as ponds silted in and streams carved through new meadows. The view from the sky corroborated what Naiman had deduced on the ground: Beavers were architects nonpareil. "Virtually every stream on the landscape was impacted in some way," Naiman told me. And those were just the *visible* changes: Nutrients and ions like nitrogen, phosphorus, calcium, and magnesium, Naiman found, collected and concentrated in meadows and ponds. Iron, for instance, increased by 118 percent. Little wonder that seventeenth-century New Englanders found the beaver's footprints so fertile.[58]

I asked Naiman to perform an act of imaginative ecology for me: to put himself in the boots of, say, Meriwether Lewis, an early explorer encountering western streams still unaltered by steel traps. What would he have found? "You didn't see nearly as much free-flowing water," he said. "A maze of beaver ponds" filled nearly every creek; even larger rivers sported beaver-built walls in their side channels and sloughs. "Rivers could move laterally onto their floodplains, which were just loaded with beavers," Naiman added. "Wide, lowland valleys would have been beaver, beaver, beaver everywhere."

———

The nineteenth century wasn't hard only on aquatic rodents — it was a bad time to be anything with feathers or fur. "Putrefying (bison) carcasses, many of them with the hide still on, lay thickly scattered over thousands of square miles of the level prairie, poisoning the air and water and offending the sight," wrote William Temple Hornaday.[59] Waterfowl were mown down with monstrous boat-mounted guns that slaughtered a hundred birds per shot. Fur seals and sea otters were pushed to within an inch of extinction to supply pelts to China. Skunks, raccoons, muskrats, deer, egrets, herons, hummingbirds — any animal whose skin or plumes could be fashioned into garb faced annihilation. The massacre was so thorough that the author Frank Graham dubbed the period the Age of Extermination.[60]

Hunting wasn't the only threat: Habitat loss took a profound toll as well. The forests went first, clear-cut for lumber, firewood, and to make room for crops. Beginning in the 1630s, William Cronon wrote, "A typical New England household consumed . . . more than an acre of forest each year."[61] By the late eighteenth century, a traveler journeying from New York to Boston passed through no more than twenty miles of wooded land, encountering little but farm fields where once forests stood.[62]

Wetlands fared even worse under their new white masters. Early colonists saw swamps, bogs, and marshes, wrote the author Rod Giblett, "as places of darkness, disease and death, horror and the uncanny, melancholy and the monstrous — in short, as black waters."[63] More than that, wetlands were wastelands, too spongy to plow or homestead. A series of nineteenth-century Swamp Land Acts deeded millions of acres of wetlands to states from Louisiana to Iowa, with the understanding that they would "reclaim" quagmires with dikes, canals, ditches, and levees. The notion that wetlands might be cradles of biodiversity didn't much bother development-minded states. In the 1600s the Lower 48 was dampened by around 220 million acres of wetlands; by the 1980s their extent had fallen by more than half, to around 100 million.[64] "So with dredge and dyke, tile and torch, we sucked the cornbelt dry, and now the wheatbelt," wrote Aldo Leopold. "Blue lake becomes green bog, green bog becomes caked mud, caked mud becomes a wheatfield."[65]

Deforestation and drainage did no favors to beavers, who, of course, depend upon wood and water for food and shelter. In turn, the elimination of our continent's original architects proved a catastrophic form of habitat destruction in its own right.

In many corners of North America, it's harder to think of an animal that *doesn't* use beaver compounds than one that does. Aquatic insects shelter in the nooks and crannies of dams and lodges.[66] Ducks nest in the grasses that spring up around pond fringes. Songbirds perch in coppicing willows.[67] Biologists have discovered that turtles and lizards are more abundant near beaver ponds in South Carolina;[68] that beavers increase plant species in streamside areas by more than a third in New York;[69] that fish communities are more diverse near beaver dams in Massachusetts;[70] that mink and raccoons hunt crawdads and snakes in beaver complexes in Wisconsin;[71] that northern leopard frogs breed in beaver ponds in Wyoming;[72] that profuse mosses grow in beaver meadows in Alaska.[73] The miracle-working rodents can even engineer habitat without building dams: A 2008 study found that wood frogs swim up castor-dug canals to reach forest feeding areas.[74] No organism is too small or too large to gain sustenance from beavers. Nutrients from castor feces breed zooplankton, sawflies lay eggs on beaver-browsed cottonwood shoots,[75] and moose are so reliably attracted to wetland plants that biologists often use the extent of beaver ponds as a surrogate for the antlered giant's range.

Add up all those dependents, and you begin to comprehend why scientists consider beavers the ultimate *keystone species*. To architects, a keystone is the wedge-shaped block that forms the apex of a stone arch, the brick that holds the span in place. To ecologists, a keystone species is that rare organism that likewise supports an entire biological community. Salmon, whose decomposing carcasses sustain grizzly bears, eagles, and even trees, are one keystone species; elephants, who clear the savanna for grasses by uprooting trees and shrubs, are another. Pull the keystone out, and the arch — or the ecosystem — collapses.

For proof of beavers' keystone status, look to the Saint Francis' satyr — a brown butterfly, about the size of a half-dollar, whose only adornment is a smattering of dark eyespots rimming its wings. A philistine like me could stroll through a field of fluttering satyrs without thinking twice. But to Nick Haddad, an energetic conservation biologist at Michigan State University, every satyr's existence is worth celebrating. That's because this unassuming insect ranks among the rarest butterflies on earth, with a population that likely numbers no more than four thousand. They are so rare that Haddad, who has spent untold hours studying *Neonympha mitchellii francisci* in the field, has seen only two wild Saint Francis' satyr caterpillars — one after a student nearly dropped his sunglasses on top of the lime-colored larva.

Most of the world's satyrs cling to life in an unlikely locale: the artillery range at Fort Bragg, a 160,000-acre North Carolina army base. When I visited Fort Bragg in 2016, it was hard to tell whether I'd stumbled upon a state park or a war zone. Golden bars of sunlight fell through the boughs of mast-straight longleaf pines, illuminating Seussian tufts of wire grass on the forest floor. The percussion of explosions and gunfire rattled the warm air, and helicopters whirred through blue haze. Armored trucks rumbled down the road, abristle with mounted guns, helmeted soldiers peeking out of the cab like gophers.

To understand why a butterfly might live in a place strafed by bombs and shells, it helps to absorb a quick lesson in satyr biology. Although no one is quite positive which plants Saint Francis' satyrs eat, they clearly favor sunny wetlands abounding with grass-like *Carex* sedges. Today that habitat is vanishingly rare in central North Carolina's forested sandhills. Only two forces open the canopy enough to permit the satyr's preferred plants. One is fire, which is why the artillery range affords prime habitat: Thanks to the constant brushfires sparked by ordnance, it's practically the only place in the state where burn cycles approach their natural frequency. (It's also a harrowing place to work: Once a year Haddad convinces the army to silence its barrages for twenty-four hours so that he can count his satyrs, escorted around the range by a demolitions expert who steers him away from unexploded munitions.)

The other force is beavers.

Although we most commonly associate the fur trade with the Northeast and the Mountain West, the Carolinas, too, once overflowed with beavers. "Those industrious animals had dammed up the water so high, that we had much ado to get over," reported English explorer William Byrd.[76] In some places Byrd's party actually had to build bridges to cross ponds. Between 1699 and 1715, however, nearly forty thousand beaver pelts departed to England from ports in Virginia and the Carolinas, creating "more wealth in the colony than indigo, cattle, hogs, lumber, and naval stores combined."[77] By 1802 John Drayton, governor of South Carolina, lamented that "the beaver is but rarely to be met with" in the state.[78] These days the best place to spot castors in the Carolinas might be at Durham's Beaver Queen Pageant, an annual gala, sponsored by the Ellerbe Creek Watershed Association, whose theme in 2017 was "Wizard of Gnawz."[79]

Few southeastern creatures were more harmed by beavers' disappearance than the Saint Francis' satyr, which relied on rodent benefactors to thin trees and create wetlands. In fact, what satyrs need, more than beavers themselves, is the years-long process of landscape evolution that beavers set in motion. When beavers first colonize a stream, Haddad told me, they often doom resident butterflies, as new ponds drown the sedges that satyr caterpillars eat. But as sediment accumulates, the pond gradually becomes a wet meadow, and the sedges return – and the butterflies with them. "There's this weird paradox where you have to accept that some butterflies are going to be killed by disturbance, but the long-term benefit is that beavers are creating habitat where the butterfly's range can increase," Haddad explained. "It's not the presence of the beaver that's important to the butterfly – it's their absence *after* their presence." The two organisms once pirouetted across much of North Carolina like partners in ancient dance, the beavers leading and the satyrs following, waltzing from wetland to wetland in intricate lockstep.

In 2011, with his beloved satyrs floating near extinction, Haddad and army biologists began trying to restart the music. In a series of ambitious restoration projects, they simulated beaver works by chain-sawing down hardwoods and plugging creeks with "AquaDams," walls of rubber tubing inflated with streamwater. Their experimental sites earned a seal of approval from the master builders themselves: Haddad happily abandoned one plot after beavers began improving his AquaDams with woody walls of their own.

The beaver mimicry, Haddad told me, was a "huge success." Some satyrs found their way to the new wetlands naturally; the team also bred the butterflies in captivity and released them into the faux-beaver sites, where the insects thrived. "We saw fifty butterflies one year, then a hundred the next year, then two hundred the next year, and now we're up to seven hundred butterflies," Haddad said. Around a quarter of all the world's Saint Francis' satyrs now dwell in the restored wetlands. Absent Haddad playing beaver, he and colleagues wrote in 2015, "it is possible that St. Francis' satyrs may have gone extinct at all sites outside of the artillery ranges."[80]

Of course, it's impractical for biologists to imitate beavers indefinitely — only the rodents themselves can create enough wetlands to truly bring back satyrs. Historically, however, beavers have not been welcome at Fort Bragg. The base is striated with dirt roads that serve as firebreaks and convey troops; when beavers flood those conduits, the army often calls in the trappers. Although Haddad and his students have constructed wire fences near crossings to prevent beavers from inundating roads, he told me the real solution may be to close some roads altogether, a measure the army is currently considering. "There's got to be some work to restore healthy beaver populations and other forms of disturbance," Haddad told me.

That other disturbance, again, is fire — and to its credit Fort Bragg operates one of the country's most naturalistic controlled burn programs. In some ways beavers are akin to fire, a maligned and misunderstood disruption that society has quelled for decades, with disastrous ecological side effects. Brainwashed by Smokey Bear's hardline anti-wildfire agenda, only recently have we woken to the importance of allowing the woods to burn. Some trees, like lodgepole pines, have serotinous cones that only release their seeds when scalded by extreme heat; birds like the black-backed woodpecker nest in the dead trunks that fires leave behind.[81] Beavers, too, create snags, by drowning trees in rising pondwaters. "As these large standing trees decay," wrote James Trefethen, "their hollows and cavities become nesting sites for squirrels, raccoons, owls, wood ducks, or goldeneyes. Their insect-infested barks and trunks become feeding places for woodpeckers, brown creepers, and nuthatches."[82] In 2001 the ecologist Mark Harmon coined the word *morticulture* to connote the importance of snags, logs, and other forms of woody detritus long ignored by foresters.[83] Beavers are adroit morticulturists: One Finland-based study found that beavers created as much deadwood as windstorms and wildfires.[84] Destruction is the preface to renewal; a force of death also breathes life.

Unlike combustion, obviously, beavers are living beings — but like fire, they're also a *process*, the catalyst for sweeping transitions that have driven the evolution of co-dependents like the Saint Francis' satyr for thousands of generations. Brock Dolman, an ecologist and Beaver Believer in California, is fond of using *beaver* as a verb — a linguistic device, he told me, meant to help his interlocutors "see the organism as an actor, as a manipulator, as an entity affecting processes over an unfolding continuum of space and time." It's a little far-out, I know, but it's also an evocative coinage. Just as a landscape can *burn*, shapeshifting from charred moonscape to grassland to timber, it can also *beaver*, from pond to wetland to meadow to forest — a centuries-long cycle of flooding and filling and growing.

At a conference in 2000, the Nobel-winning scientist Paul Crutzen first blurted the word that has come to signify our present *Homo*-dominated epoch: the *Anthropocene*.[85] The term has become shorthand for humankind's biogeophysical fingerprints, impacts that will someday fascinate and confound alien archaeologists — the radioactive isotopes deposited by nuclear weapons testing, the rapid buildup of atmospheric carbon dioxide, the buried strata of compacted straws and Poland Springs bottles and Happy Meal toys. More than a series of observable phenomena, though, the Anthropocene is an idea: that we have become the planet's principal agents of geological change, the shapers and destroyers of the modern world. "Anthropocene" isn't a happy word, but it does suggest awe — awe of ourselves.

Beavers are not entirely human-like in their environmental effects. They don't burn fossil fuels, they don't hurl waves of plastic upon beaches, they don't turn prairies into cornfields. They're motivated by food and shelter, not acquisitiveness for its own sake. But I'd submit that the difference between our two species is one of magnitude, not of kind. For beavers, too, have permanently shaped the course of biology and geology. They have re-formed rivers, kneaded meadows, filled valleys. We built our civilizations atop the sediment they left behind. As Johan Varekamp found, the signature of beavers can today be read scrawled in the seafloor. Their ponds alter our climate, storing carbon in the form of buried organic matter and releasing it as methane. They even serve, as one study put it, as "molecular geneticists": By choosing to gnaw down cottonwoods whose bark contains fewer distasteful tannins, they shape the genetic composition of riverside forests.[86] "What but the wolf's tooth whittled so fine / The fleet limbs of the antelope?" the poet Robinson Jeffers asked in "Bloody Sire," his meditation on violence and natural selection. No knock against *Canis lupus*'s dagger-like canines, but it's the beaver's orange incisors that have been evolution's most consequential dental sculptors.

Add it all up, and their impacts are continental in scale, history-changing in scope. Just as irradiated, elephant-sized cockroaches will someday scuttle through the ruins of downtown Los Angeles, so are we living in the world that beavers created. Christening a new era probably won't win me any friends among geologists, who can't even agree on when the Anthropocene began, but what the heck: Welcome to the Castorocene.

CHAPTER THREE

Deceive and Exclude

I f you were to charter a Cessna in Boston and buzz thirty miles northwest — over Medford, Woburn, and Chelmsford — you'd find yourself
flying through the airspace of Westford, Massachusetts, although, gazing
down from your cockpit, you probably wouldn't know it. Twenty thousand people live in Westford, but good luck spotting them from the air:
The town, like many New England burgs, more resembles a forest than
a metropolis. Its scatter of homes stand like islands in a sea of ponds and
woodlands. There are roads, of course, and hospitals and libraries and
community centers, but the bird's-eye view presents mostly as green and
blue, a playground for turkeys and deer.

Westford is also home to the Fletcher Granite Company, a quarry,
founded in 1881, whose stone built Boston's Quincy Market. Today most
of Fletcher's granite is sliced with diamond-edged carbide wire and embedded in street curbs. The largest hunks weigh twenty tons. Transporting
them by truck is impractical; instead, Fletcher ships its granite along
railroad tracks that run through the company's yard and connect it to the
wider world. Or at least Fletcher *did* use its railroad — until the spring of
2017, when beavers dammed an adjacent pond and submerged the tracks.
When that happened, Fletcher's people did what you do when you encounter beaver trouble in Massachusetts: They called Mike Callahan.

One damp April afternoon, Callahan pulled his dinged-up Toyota
Tundra into Fletcher Granite's yard to survey the damage. David, the
company's goateed owner, climbed in to show him around; I wedged

myself in back. We drove through a maze of sparkling granite blocks to a small wetland nearby. A cold drizzle dimpled the pond. Phragmites reeds, brown and crackling, towered in clumps ten feet tall. A distant industrial roar rose from the quarry. Beaver sign lay everywhere. Planted saplings had been whittled to toothpicks. A fresh dam blocked the outflow, raising the pond and turning the ground spongy. "You can see the top of the nest, or whatever you wanna call it," David said, gesturing to a lodge mounded on the bank. "I said to myself, 'Who cut down all the goddamn saplings?' It wasn't until I saw the chips at the bottom that I realized what'd happened."

The railroad tracks now sat below a foot of water, disappearing into the expanded pond like the wreckage of a city consumed by rising seas. "A year ago we spent a million dollars bringing those tracks up to speed," David sighed. He sounded more rueful than angry. "I kinda like the wildlife," he acknowledged. But this – this was too much.

Callahan, Fletcher's knight in shining waders, slogged along the tracks, steadying himself in the water with the handle of a rake. He is tall and wide-shouldered, with heavy features, a mop of steel-colored hair, and an easy laugh; he is the kind of avuncular personage whom younger people call, out of the blue, when they need advice. He wore a crossing-guard-orange sweatshirt emblazoned with the logo of his one-man company: Beaver Solutions LLC. Occasionally he stooped to a cut stick or bent reed, like Sherlock Holmes casing a crime scene. "It's the art of war – know your enemy," he said. "I just don't like to refer to beavers as the enemy."

"You have no fear of beavers coming up on you or anything?" David asked uneasily.

"The worst they'll do is slap their tail on the water," Callahan chuckled. Although history records one beaver-caused human fatality – a Belarusian fisherman who tried to wrangle a truculent beaver for a photo op, to which the rodent responded by severing the man's femoral artery – they're retiring creatures who, almost invariably, flee from people. "They're primarily nocturnal," Callahan said. "I almost never see them. They go to work when I'm sleeping and I go to work when they're sleeping. We're on off-shifts."

Callahan, finally satisfied, straightened to offer his assessment. "One option is to trap 'em," he said. "But nature abhors a vacuum. Beavers will come back within a year or two – and they *will* come back, because the habitat's good." Better, then, to coexist with the current tenants: to leave the beavers in place while somehow preventing their ponds from engulfing Fletcher's railroad tracks. "We need to drop this water level far enough so that your property is protected," Callahan said. The trick was to drain the pond without inciting the wrath of its occupants. "Usually I can drop the water up to a foot without the beavers being too concerned. The more I drop it, the more likely they are to try to raise the water back up."

"How long does it take them to build a new dam?" David asked.

"We could tear that out, and in a day or two they'd have it rebuilt," Callahan said. David's eyes popped. "But the thing is, you never really know what a beaver's going to do. All you can do is make guesses."

Few people are as good at guessing what beavers are going to do as Callahan, one of a handful of Beaver Believers around the country who make their living by encouraging coexistence. Although Callahan is no anti-trapping hardliner, he's convinced that letting beavers live in most cases is not only ecologically sound policy, it's also cheaper and more effective than plodding along on the trapping treadmill. Rather than eliminating delinquent beavers, he prefers to control their impacts. He waxes poetic about wrapping trees in wire, setting fences around spillways, and, most of all, regulating the dimensions of beaver ponds with pipes and fences. How can we enjoy all the good that beavers do, while thwarting all the bad?

It's a job that requires creativity, familiarity with wood- and metalworking, a solid grounding in ecological history, a great deal of manual labor, and, of course, intimate knowledge of his fur-clad combatants. It's like playing chess, except the opponent is a rodent and the goal is stalemate. "A unique skill set is required — or maybe a mindset more than a skill set," Callahan said as we drove through the rain to his next job. He thought for a moment. "It also helps to be tall."

How did beavers, once trapped to within a whisker of extinction, recover so marvelously that they're drowning quarries in suburban New England? In part, the vicissitudes of global capitalism bailed out the rodents, just as capitalism nearly destroyed them. Cheap Chinese silk found its way into the workshops of European hatmakers in the mid-nineteenth century, just as beaver pelts became prohibitively rare and expensive. Fickle fashions, along with supply and demand, shifted the market from one hat style to another. "Beaver, in days of yore, was the staple fur of the country; but, alas! the silk hat has given it its death-blow, and the star of the beaver has now probably set forever," bemoaned Robert Michael Ballantyne in 1859.[1] From the beaver's perspective, its star was at last rising again.

In the late nineteenth century, America began to snap out of the blood-lust that marked the Age of Extermination. John Muir floridly extolled the virtues of nature; George Bird Grinnell inveighed against the "evil of tree-murder" in the pages of *Forest and Stream*;[2] the young Teddy Roosevelt rallied wealthy sportsmen to the cause of conservation. The shifting zeitgeist found expression in legislation as well as at New York supper clubs. Yellowstone National Park, the nation's first, was designated in 1872. States passed laws protecting fish, game, and fowl, which, in 1900, received federal backing via the Lacey Act, a law that illegalized transporting poached wildlife across state lines. Roosevelt, upon assuming the presidency, racked up conservation accomplishments, creating the National Wildlife Refuge System in 1903 and the US Forest Service in 1905. Protected from hunting and habitat loss, persecuted species — black bear, moose, Canada geese — began to reclaim their former haunts.

Although some writers, like Muir, advocated defending nature for its own sake, the high priests of conservation mostly preached a utilitarian gospel. "Conservation," Forest Service chief Gifford Pinchot wrote, "means the wise use of the earth and its resources for the lasting good of men."[3] And what could be more useful than a beaver? The once-heretical tenet that beavers had more value as engineers than pelts became a minor dogma in the new religion.

The most ardent evangelist was Enos Mills, a perspicacious naturalist and slightly maudlin nature writer who communed with beavers for years near his mountain home in Estes Park, Colorado. One memorable autumn, Mills visited the same lodge daily for sixty-four consecutive days, scribbling meticulous notes about the dimensions of dams and canals and detailing the activities of the resident colony. "Beaver have great fun while growing up," he observed with typical high spirits. "In the water they send a thousand merry ripples to shore."[4] In 1913 Mills summarized a quarter century of study in *In Beaver World*, an adoring profile of his dam-building neighbors. "The beaver is practical, peaceful, and industrious," Mills cheered, adding, rather judgmentally, "These and other commendable characteristics give him a place of honor among the hordes of homeless, hand-to-mouth folk of the wild."[5] Through close and persistent study, Mills intuited many of the ecological benefits that contemporary scientists are still attempting to quantify. The beaver, he wrote, was the "original Conservationist," responsible for checking erosion, blunting floods, keeping streams perennial, and bolstering trout. "What he has accomplished is not only monumental," Mills added, "but useful to man."[6]

Posterity has largely forgotten Mills's influence: His writing never quite achieved Muir's ecstatic heights, Thoreau's depth, or Leopold's erudition. In his day, though, he was as significant as any peer; one profiler wrote that Mills had "done more than any other man in the West to bring nature and human kind together."[7] Beavers remained a lifelong cause: One delightful photograph depicts Mills — a corona of frizzy hair encircling his pate and a chewed branch in his hands — leading a troop of several dozen librarians, primly attired in dresses and suits, on a tour of his beloved ponds.[8] He brought gangs of city children to Estes Park, where he enjoined them to sit for "hours upon a log by a beaver pond."[9] Mills wasn't the first to remark upon beavers' salutary power, but he had the widest audience. "A live beaver," he concluded, "is more valuable to mankind than a dead one. . . . May his tribe increase!"[10]

Natural resource agencies around the country were equally sympathetic. Vermont, Utah, California, and a host of other states reintroduced beaver to their lakes and streams. The most spectacular success came in New York, whose population had, by the 1890s, dwindled to a single five-animal colony clinging to life in the Adirondacks, the forested massif that sprawls across the state's northeast corner.[11] Between 1901 and 1907, officials relocated thirty-four beavers — some trapped in Ontario, some in Yellowstone, and some acquired from Canadians at the 1904 Louisiana Purchase Exposition in St. Louis — into Adirondack Park.[12] Protected by law from trapping and blessed with unlimited access to water and wood, the critters flourished. By 1915 New York's population had erupted to fifteen thousand beavers, and in 1923 the state deemed beavers healthy enough for trappers to resume their pursuit of pelts.[13]

The art of beaver relocation found expression on the opposite coast as well. In 1933, with America mired in the Depression, President Franklin Delano Roosevelt founded the Civilian Conservation Corps, a program that put hale young people to work building trails, planting trees, fighting fires, and laboring at other outdoor jobs. Among the corps's responsibilities was tackling soil erosion, a natural phenomenon that, as the Dust Bowl revealed, was being accelerated by benighted land use. Alongside its human erosion control agents, the corps released around six hundred beavers into streams in California, Wyoming, Oregon, and Utah. The rodents represented one of the New Deal's best bargains: They cost the feds five dollars apiece to capture and plant, and provided soil services later valued at three hundred dollars per animal.[14] North Dakota and Washington attempted relocations as well, though they didn't document their success rates. California went the wildest: Between 1923 and 1950 officials introduced 1,221 beavers. Many of the Golden State's modern colonies descend from those intrepid transplantees.[15]

Thanks to reintroduction, conservation laws, and benign neglect, beavers roared back. In 1928 the Adirondack population spilled into Massachusetts, establishing the Bay State's first colonies since the 1770s.[16] By 1955 beavers had recaptured every inch of New York beyond Long Island and New York City. A brief article in the Ohio Academy of Sciences journal noted an appearance in Cincinnati in 1985 — perhaps the first since 1755.[17] Gnawed willows toppled again in Louisiana's bayous.[18] From the marshes of Maryland to the coasts of Oregon, the beaver tide began to rise.

At first folks of all stripes celebrated the comeback. Conservationists got their ecosystem engineers, trappers a renewed stock of pelts. But it wasn't long before the resurgent rodents caused trouble. No sooner had they crossed the border into Massachusetts than they irked a farmer by flooding his road and crops.[19] That early clash proved a harbinger of conflicts to come. The landscapes to which beavers returned bore little resemblance to the places they'd once thrived; like Rip Van Winkle, they found their environment transmogrified in their absence. Beavers had disappeared from the Northeast during the age of the horse and buggy. While they were gone, as Jim Sterba put it in *Nature Wars*, "much of the vast grid that keeps America moving and humming was built on prime beaver habitat."[20] Power lines, telephone wires, railroad tracks, highways, and suburban sprawl proliferated where construction was easiest — in the river valleys and lowlands beavers once called home. Roads bifurcated wetlands; farms stood next to ponds; houses popped up everywhere. Never before had so many beavers and so many humans shared such tight quarters. Our perception of beavers shifted as we clashed, yet we remained antagonists, our interactions still mediated by the jaws of traps.

Ed Grohoski is among the many landowners who have butted heads with the exasperating mammals. Grohoski lives in eastern Maine, on a ten-acre property that consists largely of swamp and forest — sugar maple, beech, hornbeam, ash. His home is heated by a fireplace and a woodstove, which he fuels with logs he cuts himself. He takes pride in stewarding his forest and the otters, moose, deer, owls, foxes, and skunks who call it home. "The one thing I don't encourage is wild turkeys," he told me. "My dogs have taken to rolling in their dung."

But not even the gobblers drive Grohoski as mad as beavers. Although the rodents were scarce when he purchased his lot in 1982, they soon showed up in force; Grohoski realized he had a problem when a six-hundred-foot-long dam so swelled his swamp that it cut off access to his back acres. The beavers felled dozens of trees, threatening his fuelwood supply. Grohoski didn't despise his beavers — as a retired electrician, he harbored a professional admiration for their craftsmanship. He also found their dams furnished useful footpaths through his swamp, as tribes like the Penobscot discovered thousands of years earlier. But he couldn't tolerate their takeover.

"To have their water features there is an asset to me and all the creatures that use the habitat," he said. "To have the water there in such total abundance where I can't even get to the back of my property — that's not in my best interest."

Grohoski did what most landowners do: He hired a trapper, who extracted six beavers from the property in the winter of 2014. The respite wouldn't last. By 2016 beavers had recolonized the swamp, likely swimming up to Grohoski's lot from nearby Branch Lake. The trapper came back and claimed another half a dozen. This time the reprieve was even briefer: When I talked to Grohoski in September 2017, he told me he'd lately spotted another beaver — judging from her size, probably a dispersing two-year-old.

I asked Grohoski if he expected to continue trapping beavers as long as he owned the lot. "Absolutely," he said. "They'll win eventually, because I'm going to die. Nature always wins."

Ed Grohoski isn't the first person to learn how inexorable beavers can be. In 1984 and '85, Allan Houston, then a researcher with the University of Tennessee's Agricultural Experiment Station, trapped every single beaver on the Ames Plantation, a stream-laced property fifty miles west of where the Mississippi River churns through Memphis. Then he monitored the plantation for the next forty months, removing any luckless beaver who attempted to claim the unoccupied territories.

Houston's results provided a striking illustration of rodent resilience. He eliminated 169 resident beavers in his first wave of trapping — and 162 would-be colonists over the next forty months. The creatures were Hydra-like in their relentlessness: Every beaver Houston slayed, it seemed, was immediately replaced by another set of hungry incisors. He didn't miss the implications of his findings. "Potential immigration into these domains makes it probable that control programs," Houston wrote, ". . . will be as perpetual as the resource they are designed to protect."[21]

———

As beavers returned to eastern landscapes, nowhere did they inflame passions more than in Massachusetts. By the 1980s the state's beaver population had crept up to twenty thousand, and damage reports were escalating. The state's Division of Fisheries and Wildlife, or MassWildlife, began to fear that beavers were approaching their *cultural carrying capacity*: the number of beavers — or black bears, or deer, or any other potentially irritating wild animal — that humans can tolerate. Cultural carrying capacity is the squishier sibling of *biological carrying capacity*, the number of animals that a given habitat can sustain. While biological carrying capacity is, in theory, a hard-and-fast threshold — there are, after all, only so many miles of open stream in New England — cultural carrying capacity is harder to pin down, more dependent on societal attitudes and public perception than physical reality. A populace that's convinced that beavers are pests will inevitably have a lower capacity than one that considers them a habitat-creating boon.

Whatever the logic, MassWildlife decided that twenty thousand was about the right number of beavers, and vowed to maintain that level. Lacking predators, the state claimed, beavers would multiply like bacteria in a petri dish. To forestall this apparently nightmarish scenario, MassWildlife turned beaver control over to fur trappers, who paid the state for licenses, reported their catch, and sold the pelts.

But it wasn't long before the trapping program came under attack. In 1996 the Humane Society of the United States and other animal welfare groups sponsored a ballot initiative, called the Massachusetts Wildlife Protection Act, or "Question 1," that would outlaw supposedly cruel leg-hold traps and a body-gripping model called the Conibear. The Conibear's history is drenched in irony: It was the invention of Frank Conibear, a Canadian trapper who'd designed the apparatus to crush its victims like a giant mousetrap, averting the prolonged agony inflicted by a leg-hold. In 1961 he even received the American Humane Association's first Certificate of Merit.[22] But Conibears don't always kill instantly, and activists fretted that beavers who drowned in the traps would suffer for many long minutes before expiring.[23]

The campaign turned nasty, in the way that animal rights battles often do. The Humane Society's coalition aired TV ads depicting pet dogs squirming in traps. MassWildlife issued a press release condemning the initiative, and was promptly reprimanded for using state funds to lobby voters. *The Boston Globe* editorialized that the trap ban was "misguided." Even so, on November 5, 1996, the initiative passed. Leg-holds and Conibears were generally off the table. You could still kill beavers, to be sure — some folks continued to catch them in live traps and shoot them, a method that was arguably less humane than Conibears — but many recreational trappers quit chasing castorid quarry, incensed at losing their favorite tools.[24]

MassWildlife officials issued apocalyptic warnings about the coming beaver tsunami. "It wouldn't surprise me if [the beaver population] will be about 100,000 in five or seven years," one cautioned in 1998. "They're rodents. They're highly productive."[25] By 2001, the state claimed, the population had spiked to around seventy thousand — more than triple its size before the trap ban. Annual complaints rose from four hundred to a thousand.[26] (Dave Wattles, the state's furbearer biologist, told me the ban had ended MassWildlife's ability to estimate population size, but that beaver populations today appear to have bumped against habitat limits and stabilized.) Local newspapers threw logs into the story's fire. "Trap Ban Gives Beavers the Run of Rural Town," blared one headline, conjuring a rowdy beaver posse brawling in a dusty saloon. "Trap Law Leaves State

Knee-Deep in Wildlife," wailed another, as though no fate could be more awful than being surrounded by animals.[27]

———

At the time the fracas erupted, Mike Callahan's life couldn't have been less entangled with beavers. Callahan worked as a physician's assistant at a methadone clinic in western Massachusetts, trying to help opiate addicts stay clean. His wife, Ruth, was a nurse. Together they ran a bed-and-breakfast. Theirs was a peaceful life, unmarred by beavers or controversy. When they encountered the rodents on camping trips, Callahan thought they were kind of cool, and nothing more.

Gradually, however, beavers dragged the Callahans into their orbit. The couple kept a close eye on the news and supported the 1996 ballot initiative, growing ever more dumbfounded by the vitriol that swirled around the retiring rodents. They learned about beavers' role as a keystone species, attended workshops sponsored by animal protection groups, and traveled to upstate New York for a beaver conference. Nonlethal beaver management became an interest, then an obsession. "We thought there had to be a middle ground," Callahan recalled to me.

He and Ruth were most intrigued by the potential of *flow devices* — pipe-and-fence systems that partially drain ponds by creating a leak not even a beaver can plug. Although the term serves as an umbrella for any manner of contraptions, all flow devices function through a combination, as one reviewer put it, of "deception and exclusion."[28] Picture, for example, a python-length plastic pipe — its mouth submerged in a beaver pond, its body passing through the woody bulwark of a new dam, and its tail end gushing water into the creek downstream. That's the deception: The perplexed beavers, conditioned by evolution to fortify their dams against leaking, fail to grasp that water is actually departing the pond through the pipe. Both pipe ends, meanwhile, are surrounded by metal cages that prevent beavers from detecting and plugging the leak. That's the exclusion. The frustrated rodents may go mad reinforcing their dams, but they're powerless to prevent the water's escape. When all goes well, beavers eventually give up, accepting their diminished pond as the new status quo.

Maybe that all makes a straightforward technology sound more complicated than it is, but here's the gist: If you enjoy having beavers in your backyard but are not particularly keen on having to snorkel through your basement, you can install a flow device — or, better yet, have a professional install one — to prevent the wetland from inundating your home.

Entranced by the dual power of deception and exclusion, Mike and Ruth convened a volunteer group of flow device installers to tackle conflicts themselves. Callahan traveled to Canada to meet Michel Leclair, a former trapper installing flow devices in Ottawa, and organized a workshop starring the Vermont-based Skip Lisle, the world's foremost castorid conflict mediator. Early in their training, the Callahans caught wind of a meeting in nearby Amherst, where town leaders would decide how to manage beavers near a popular bike path. The couple, worried they weren't ready

Flow devices can prevent beavers from plugging culverts and washing out roads. The pipe passes water through the culvert, while fences deter beavers from clogging the pipes' ends, saving transportation departments money and sparing beavers' lives. *Illustration by Sarah Gilman.*

to tackle such a high-profile trail, agreed to sit in back, take notes, and inconspicuously get a feel for local politics. "So we go to the meeting, walk in, and there's no one in the room except for the town board," Callahan recounted. "They turn to us and say, 'You must be the beaver people.' We spent the next hour trying to convince them we didn't know anything, and they spent the hour trying to convince us to do it."

The council proved persuasive. The Callahans protected the bike path with a flow device, and Massachusetts's beaver-crazy media duly covered the story. Requests for new devices began trickling in. The volunteer group scaled up. Callahan found he enjoyed beaver work more than his day job as a physician's assistant. "It was immediate gratification," he told me. "With the opiate-addicted people I was working with at the methadone clinic, a lot of them get better, but it's a long process, and it doesn't always succeed like you would hope. Whereas working with beavers — it's like, *boom*, in one day you can see the difference."

In 2000 he converted the volunteer group into a for-profit business, Beaver Solutions. By the time I visited him in May 2017, he'd installed more than thirteen hundred flow devices, most of them variations on designs originated by Leclair and Lisle, the giants upon whose shoulders he stood. Callahan had recently hired a developer to redesign his website, and his new Google-friendly page was funneling him more work than he could handle. "Right now I've got sixty-one jobs at different stages," he told me the day we spent driving around Westford. "Not to mention 450 sites that I have to do spring maintenance checks on. And the new requests just keep coming in."

———

Callahan's most loyal customers aren't homeowners, but transportation agencies: The Massachusetts Department of Transportation has had him on statewide contract for over a decade, and local highway departments account for more than half his business. More, even, than wildlife agencies, road crews determine the fate of northeastern beavers. The reason is one of America's most vulnerable, most ubiquitous structures: the humble culvert.

Culverts, for the uninitiated, are those big pipes — made of corrugated metal, concrete, or plastic — that funnel streams and wetlands under railroad tracks and roads. They are essential: Without them, many thoroughfares would flood after every rain shower. They are everywhere: One biologist has estimated a quarter-million culverts gird New England alone.[29] And until culverts fail — and they *do* fail — they are invisible: You have walked or driven over hundreds in your life, although you may not have realized until this very moment that these nondescript pipes even had a name.

Culverts are Achilles' heels in the corpus of American infrastructure. They are the friction points at which the natural environment rubs against the built one, fragile joints worn thin by the rush of water. Occasionally they erode and buckle, collapsing the roadbed above. Sometimes they're swept away by heavy flows. Mostly, though, they clog. Storms flush logs and sticks downriver, plugging culverts with gnarled nests of woody debris. Unable to find release through the jammed pipe, the rising stream spills onto the roadway, sometimes washing it out altogether. When Hurricane Irene tore through the Northeast in 2011, its floods destroyed two thousand road segments and damaged more than a thousand culverts. And the only force that's harder on culverts than hurricanes is beavers.

Imagine briefly that you're a beaver — a dispersing two-year-old male, say. You've recently departed your lodge, supplanted by newborn siblings who have become the apples of your parents' beady black eyes. You're house hunting. You have to find deep water or build a dam soon — you can smell the funk of nearby black bears — but the best homes are taken. You acquire a notch in your tail attempting a coup against a larger male; you escape with your life, but now you're getting frantic. You're swimming downriver one evening, wondering where you'll spend the night, when, miracle of miracles, you come upon the finest dam you've ever seen. It's a huge, immovable wall of rock and earth, set perpendicular to the stream — the bank of the human-built road that passes over your creek. There's only one chink in the armor of this wondrous structure: a circular metal pipe that steers your stream through the berm and out the other side. A culvert. Plugging the culvert with sticks proves a mere night's work. The upstream side swells into a pond, and you're home safe. The roadbank is your dam. The culvert is the leak. And you hate leaks.

Just as beavers are hell on culverts, culverts are hell on beavers. Transportation departments nearly always handle castorid-clogged culverts by calling the trapper. Culverts become rodent honeypots, luring beavers to their death with the promise of prime habitat. "Road crews continually seem to believe that the best remedy is killing beavers, despite the massive and ongoing evidence to the contrary," Skip Lisle told me. "Evidence in the form of having to clean out the same culverts over and over again over the years."

One late-summer day, on a muggy morning promising afternoon rain, I drove down a tortuous country lane in a lush corner of Vermont to pay Lisle a visit. Like many rural mill towns, Grafton has been downsized by a centuries-long human exodus. More people lived in this part of Vermont in the early nineteenth century than linger today, and the brown clapboard house Lisle shares with his wife, Elise, is the only human habitation for a quarter mile. What Lisle's neighborhood lacks in people, it makes up in aquatic rodents. As I jounced toward his home in my Camry, the rocky track emerged from oak and maple forest to cleave a sunlit wetland terraced with overgrown dams. Two lodges, one marked by a fluttering American flag, punctuated the pond. This is the kingdom that Beaver Deceivers built.

Lisle, founder and owner of Beaver Deceivers International, met me in the driveway, an exuberant black Lab named Cally on his heels. Lisle cut an impressive figure, his considerable height and Rock Hudson jawline made more imposing by his upright carriage, trim gray hair, and tucked-in polo shirt. He gave off the vibe of a high school wrestling coach, all discipline and good health.

His home, the same one in which he'd grown up, was a cheerful clutter of *Smithsonian* magazines and Will Ferrell movies, along with the requisite wooden beaver figurine. Lisle removed the shade from a living room lamp, revealing a bulb fixed in a block of beaver-chewed aspen. His parents, he told me, had been furniture-makers whose pieces often incorporated the rodents' work. Some of Lisle's earliest memories involved driving around at dusk with his folks, hunting for promising wetlands. "They were fascinated by the things that beavers did," he said, "even though they wanted 'em dead when they cut down their own trees." In 1970, when Lisle was twelve years old, his parents demanded he use his bolt-action .22 Mossberg to gun down a beaver near their pond. He obeyed, reluctantly. "That was the last beaver I ever killed," he told me.

Several years later a beaver dammed the culvert beneath the road that ran through his family's property — the same low-lying road I'd driven to reach the house, and an exemplar of ill-conceived urban planning. Lisle commandeered a few lengths of his dad's garden fence, bent it into a crude cage, and affixed it just upstream of the culvert. The water could enter the culvert, but the beavers couldn't. The slapdash construction was the first Beaver Deceiver, a deceptively simple flow device that Lisle has spent the rest of his life refining, disseminating, and defending.

We walked down to the Lisle family wetland, Cally in pursuit. When his parents were alive, Lisle said, they'd religiously mowed the yard. After Skip inherited the house and moved back in 2001, he'd let it go; now unruly jumbles of junipers, ferns, and blackberries braided the property, and wild apple trees, the long-buried legacy of a colonial orchard, drooped with fruit. A woodpecker jackhammered a snag. "I love dead trees," Lisle sighed happily. I had never seen so many amphibians in one place. Wood frogs, green frogs, and pickerel frogs plooshed into the water at our approach; peepers bounded between logs; red efts skulked in wet leaves. The air trembled with dragonflies; the brush was alive with garter snakes. A row of painted turtles plunked, one by one, off a dead log. The pond's surface was carpeted in an emerald skein of watershield, a perennial lily-like plant that the beavers seemed to consider a delicacy. By creating a still, sunlit, watershield-friendly pond, Lisle said, beavers had essentially cultivated their own food supply.

Lisle spread his arms to encompass the abundant scene, feeling, perhaps, like Noah surveying the passengers aboard his ark. "By preventing us from having to shoot beavers, that original Beaver Deceiver has been responsible for producing millions of life-forms," he said. "Not every site lends itself to this, but thousands do. That's the incredible ecological potential that beavers have."

If the field of flow device installation were theoretical physics, Skip Lisle would be its Isaac Newton. After receiving a master's degree in wildlife management at the University of Maine, Lisle worked for the Penobscot Nation, a native tribe that owns 150,000 acres in wet, woodsy, beaver-filled central Maine. Lisle installed dozens of Beaver Deceivers at roadside conflict points around the Nation in the late 1990s, sometimes attracting curious stares from tribal members who, as Lisle put it, "seemed to be betting on the beavers." By the time he moved on more than six years later, Penobscot Nation had become, he told me, "the first large landowner anywhere in the world to completely beaver-proof their land nonlethally."

In 2001 Lisle went into business for himself, contracting with public agencies, private landowners, and nonprofit groups like trails associations — just about anyone plagued by beaver flooding. He also conducted workshops and trainings, schooling Mike Callahan and others in the finer points of deception and exclusion. Lisle's Beaver Deceiver design evolved, morphing into a wood-framed trapezoidal fence whose short side he placed by the culvert's mouth. Beavers began their dams close to the culvert, at the trapezoid's narrowest point; as the rodents frantically laid sticks along the fence, the trapezoid's angled sides directed them away from the culvert. The more the beavers dammed, in other words, the farther they found themselves from the leak.

As the years went by, Lisle designed a blizzard of colorfully named inventions. There was the Castor Master, a pipe system that passed water through culverts or freestanding dams; the Misery Multiplier, a double layer of fencing that deterred even young beavers from squeezing through; and the Round Fence, which is, well, a round fence. He also perfected his signature innovation. Around 2000 Lisle abandoned simple trapezoidal Beaver Deceivers in favor of custom-built designs that exploit each site's unique topography. Although he holds a patent on *Beaver Deceiver*, I've heard the term used to describe, incorrectly, just about any device meant to outsmart a castorid, from a flimsy length of PVC to a bale of chicken wire wrapped around a cottonwood. "It's taken off like Kleenex," Lisle sniffed.

One town that has no misconceptions about Beaver Deceivers is Andover, a hamlet of twenty-three hundred smack in central New Hampshire's Lakes Region. Vicky Mishcon, the charming, white-haired chair of the town's board of selectmen, told me Andover had forever been plagued by culvert blockages and washed-out roads. The town traditionally handled those problems by hiring a contractor to ram out the debris with heavy machinery, at a cost of $150 per hour, and a trapper to kill the offending beavers, at $100 per pelt. "Of course, then you haven't done anything but put up a vacancy sign for a new family," Mishcon said. In 2004 a downpour swept over a clogged culvert and washed out a throughway called Elbow Pond Road, requiring a forty-eighty-thousand-dollar repair. No sooner had the town fixed the road than beavers plugged the culverts again. "Everyone was kind of scratching their heads," Mishcon said.

The town's conservation commission approached Lisle about installing a Beaver Deceiver to protect the vexing culvert. The device, Lisle told them, would cost around three thousand dollars. "People in town did *not* want to spend the money," Mishcon said. "They said that it was a waste, that it wouldn't work." Andover hired Lisle anyway. "Skip starts attracting attention," Mishcon recalled. "He's this big guy, he's got a truck full of posts and steel fencing, and he's building these odd-shaped trapezoidal things. People are stopping by, he's chatting with them, and soon he's got the whole neighborhood down there." Lisle built one Beaver Deceiver in Andover, and then another, and another — around ten altogether. Fish, fowl, and frogs flourished; the road washouts ended. "We now have a line item in our budget for maintenance and new Beaver Deceiver installations," Mishcon told me. "People don't complain anymore. It's been such a saver of time and money."

Buoyed by his success in Andover and elsewhere, Lisle waited for New England's wildlife agencies and transportation departments to come knocking. And waited. "The flow device thing seemed so obvious to me that I naively thought I could change the face of beaver management," Lisle told me. "And now here I am twenty years later, and we seem to be going backward."

Many agencies, Lisle found over the years, didn't just lack interest in flow devices — they actively insisted the contraptions didn't work. A paper published by MassWildlife scientists in 1997, for instance, alleged that flow devices could solve less than 5 percent of the state's beaver conflicts, a claim that many media outlets trumpeted during the Question 1 wars.[30] As recently as 2016, New Hampshire's furbearer biologist told NPR that flow devices were ineffective and expensive.[31]

Vicky Mishcon, the Andover selectwoman, witnessed that recalcitrance firsthand when she pitched New Hampshire officials on beaver-proofing a state-owned road near Elbow Pond. "We all sat around the table, presented our argument, and asked if they would hire Skip," Mishcon told me. "This guy sat there and at the end of the meeting said, no, they're not interested, these flow devices don't work, and the permitting process was too difficult. I was so angry. How can you say they don't work — you've been in our town, you've seen them!"

The federal government isn't much more enthusiastic. Wildlife Services, the branch of the US Department of Agriculture responsible for managing troublesome animals, killed 21,184 beavers in 2016 alone.[32] In one typical environmental analysis justifying beaver-killing in Virginia, Wildlife Services claimed that flow devices "may be ineffective in beaver ponds in broad, low-lying areas . . . may not be appropriate in streams or ditches with continuous flow . . . [and] may not be effective during periods of unusually high rainfall or water flow."[33] A reader could be forgiven for wondering when, exactly, flow devices *do* work.

Jimmy Taylor of the National Wildlife Research Center, the agency's research arm, told me that Wildlife Services and its collaborators use flow devices in some states, but that many acute conflicts call for trapping. "If transportation departments have flooding over a public road, they need that water gone *yesterday*," Taylor said. "They might be interested in a flow device long-term, but they have to do something right now to protect human health and safety."

As Taylor rightly pointed out, scientific literature documenting the effectiveness of flow devices is sparse. Some papers have cast doubt on their efficacy: One 2000 Mississippi study of Clemson pond levelers – an early design that's largely fallen out of favor – found that half the devices failed.[34] More recent studies, though, suggest the devices are worth it, and then some. In 2008 researchers Stephanie Boyles and Barbara Savitzky tallied up the costs and benefits of flow devices that Skip Lisle constructed in coastal Virginia. Before Lisle's intervention, the Virginia Department of Transportation shelled out a whopping three hundred thousand dollars a year on maintenance, repairs, and trapping at fourteen conflict-prone roads around the state. The agency's beaver control approach was so inefficient that it was underwater, literally and financially: Every dollar the state spent on beaver management and repairs saved just thirty-nine cents.

After Lisle installed thirty-three flow devices at a cost of forty-four thousand dollars, beaver damage instantly ceased. The cost-benefit ratio of Lisle's contraptions was stunning: For every dollar the Department of Transportation paid the Beaver Deceivers maven, it earned more than eight dollars in savings.[35] A 2017 Alberta-based study by the beaver researcher Glynnis Hood documented similar gains, finding that a dozen flow devices installed in a wetland park near Edmonton could save the provincial government around $180,000 over seven years.[36]

Mike Callahan has chronicled equally impressive benefits. In a non-peer-reviewed paper published in the journal of the Association of Massachusetts Wetlands Scientists, Callahan found that his culvert-protecting flow devices succeeded 97 percent of the time, and that pond-lowering devices installed through freestanding dams were 87 percent effective.[37] The costs, annualized over a decade, ran between $200 and $290 per year — a fraction the price of road repairs, and around $14 for every acre of beaver-built wetland saved by forgoing trapping.[38] (For comparison's sake, the average cost of a wetland restoration project in 2006 was a whopping $38,275 per acre.[39]) Flow devices look good even if you only care about protecting roads and culverts — and once you start tabulating the value of habitat preservation, the fight becomes a rout.

Even so, when Callahan launched Beaver Solutions, even his own brother assured him that flow devices didn't work. "There's a huge need for an updated study on cost-effectiveness," Callahan acknowledged to me. In 2017 he founded a group called the Beaver Institute, a nonprofit counterpart to Beaver Solutions, which will fund such research, train flow device installers, help landowners afflicted by conflicts, and promote beaver education. He isn't against lethal control — he recommends trapping at around a quarter of his sites — but the reflexiveness with which many wildlife managers turn to trapping is one of the few subjects capable of darkening his sunny disposition. "This is their field!" he railed when we met in Massachusetts. "You'd think they'd be up on the latest and greatest information. But they're just spewing a load of outdated crap." Claiming beavers as your national mammal doesn't guarantee smarter management: Glynnis Hood found that around three-quarters of Alberta's municipalities and park districts handle beaver conflicts with guns and traps, while just 5 percent employ flow devices.[40]

Ask Skip Lisle, and he'll tell you that the problem is an asymmetric burden of proof. Many state wildlife managers demand ironclad evidence that flow devices work, but their standards aren't nearly so high when they prescribe trapping. "No one is ever skeptical about killing beavers and cleaning culverts in an endless cycle," he said as we meandered around his wetland. "When I try to convince them to hire me, I say, look, what you're doing now is not working — that should be clear. But there are a great number of people at a great number of agencies who have come to the conclusion that flow devices don't work and beavers will always outsmart them."

One exception is Lisle's home state, Vermont, whose Department of Fish and Wildlife has installed around two hundred and forty flow devices since 2000 at a success rate that has recently approached 90 percent. "I think we've made a lot of progress in increasing people's support for these animals," Kim Royar, the state's furbearer biologist, told me later. Royar is as pro-castor as state biologists come: Unlike other managers, she rejects the term *nuisance beaver*, on the grounds that it's humans who cause the conflicts. Still, her agency's efforts are tempered by caution. Royar told me the state avoids installing its "Beaver Baffles" in fast-flowing water and high-sediment streams. That skittishness means Vermont only handles a small fraction of beaver complaints with a flow device — the lowest-hanging fruit. Royar said the state occasionally refers cases that surpass its own expertise to Skip Lisle. But many landowners reject the help, and many beavers still die.

As Lisle and I emerged from behind a stand of beech, a quintet of dabbling wood ducks exploded into flight, splashing down moments later at a safe distance to fluff their feathers in annoyance. Lisle gazed out at the pond's twin lodges, their frames spangled with light-loving flora that flourish in open beaver flowages — wine-dark clusters of elderberry, purple bouquets of joe-pye weed, delicate orange vases of touch-me-not. "I do not understand people generally," he said, as Cally disported at his feet. "Beavers are so simple. I never get angry at a beaver, because they never act maliciously. Whereas our species is very prone to acting that way."

———

In the years since Mike Callahan learned flow device installation under Skip Lisle's tutelage, their paths, like those of many erstwhile mentors and mentees — Anakin and Obi-Wan, Batman and Ra's al Ghul — have diverged. Most obviously, while Callahan considers some last-resort trapping unavoidable, Lisle has an almost Jainist aversion to killing beavers. Their approaches to sharing knowledge differ, too. Callahan has distributed around two thousand copies of a self-help DVD designed to provide landowners a crash course in flow devices. Meanwhile, do-it-yourself installations are the bane of Lisle's existence.

"It's easy for people to look at what I do and say, oh yeah, anybody can think of that," Lisle told me. "But it's taken a great deal of time and struggle to perfect these flow devices. I'm as experienced and successful as anybody out there, and yet managing conflict points is still an immense technical challenge for me. That's why I resist the idea that anybody can do this if you just get the word out. You have to be able to look at other people's work and say, this is a piece of crap. You might offend them, but you can't solve the problem without being honest about it."

It's hard to argue he's wrong. I've spoken with numerous landowners and managers who claim to have tried and abandoned "Beaver Deceivers." When I've asked them why their devices failed, I'll invariably learn that their DIY contraptions were doomed by obvious design flaws. And when well-intentioned but untrained laypeople gossip with their neighbors, the greater cause of coexistence can suffer. "One of the Forest Service guys around here put in something he called a 'Beaver Deceiver,'" Mike Settell, director of a scrappy Idaho nonprofit called Watershed Guardians, told me. "It was just a little pipe with a bunch of holes in it, and it plugged up, and he told everybody that Beaver Deceivers don't work. Well, of course *that* crappy thing doesn't work."

Until we train an entire army of Callahans and Lisles, though, North America will remain plagued by beaver conflicts. In that light, Callahan's egalitarianism seems like the only way to beaver-proof the country. "We need both Mike and Skip," one Beaver Believer told me. "We need someone who says, Look, this isn't rocket science — and we need someone who says, This *is* rocket science, and I am a rocket builder."

In the afternoon the rocket builder took me along to observe some rocket maintenance. Although Lisle had effectively beaver-proofed all of Grafton's vulnerable roads, insurgencies occasionally erupted. The crafty rodents had recently penetrated a Beaver Deceiver on a neighbor's property, clogging a culvert and threatening a forest road. "Fucking beavers, is all I have to say," Lisle quipped as we surveyed the daunting berm of leaves, sticks, and muck in which the indefatigable builders had encased both culvert and flow device. "It's a good kind of anger, though," he added. "I know they're not doing it to be mean to me."

Lisle yanked on a pair of chest waders, grabbed a pitchfork from the bed of his truck (whose bumper came complete with BDI vanity plate), and lowered himself into the pond. He attacked the clot ferociously, baling away basketball-sized clods of mud, biceps straining under his sleeves. I stood uselessly in the waist-deep pond, admiring the chain of dams and pools receding upstream into the red maples.

Lisle's goal, he told me, was to make himself obsolete, to mediate so many culvert conflicts that the thorny problem of coexistence evaporated.

Wildlife biologists often talk about cultural carrying capacity as though it's an immutable number. But the only reason we struggle to tolerate beavers is that we're so bad at preventing them from drowning our roads and property. The better we get at averting conflicts, the more beavers society can tolerate — and the more fully we can reap their rewards. "Most people don't care about beavers being on the landscape as long as roads aren't being flooded," Lisle grunted as he forked out another wad of gunk. "Once you protect the culverts, the cultural carrying capacity becomes basically whatever the landscape will allow."

A drizzle began to tap dance on the wet leaves, and soon Lisle was soaked through — whether with rain or sweat, I couldn't tell. One pitchfork-load at a time, he tore apart the clog. Soon a small whirlpool formed near the culvert, as though he'd unplugged a bathtub drain. The pond's level plummeted as its contents escaped beneath the road; Lisle tossed fistfuls of sticks and leaves into the rejuvenated flow, letting the stream sweep away the disassembled dam. Soon the pond was once more a creek, exposing the submerged Beaver Deceiver within: a handsome, durable-looking structure of yellow pine and epoxied steel that fit the stream's contour so snugly it seemed nearly natural.

Altogether, the maintenance had taken no more than forty-five minutes. (Mike Callahan has estimated that his devices require just an hour of upkeep a year.) Lisle resolved to affix another sheet of fencing atop the Deceiver. We slogged back to the truck, pitchfork slung over Lisle's shoulder, Cally weaving around his knees, the rain hissing through white oaks.

"The idea that we can't make flow devices work is almost embarrassing," Lisle said. He tossed the pitchfork in the truckbed and turned to me. "Beavers have a brain the size of a walnut.[41] If we want to, we can beaver-proof the freakin' world."

— CHAPTER FOUR —

The Beaver
Whisperer

When Sandy and Chomper met in their concrete-floored pen in Winthrop, Washington, they found love at first sniff.

He was a forty-four-pound male with quick, inquisitive eyes who'd been busted for gnawing down apple trees. She was a twenty-three-pound female, coated in lustrous ruddy fur that glinted blond in the sun, incarcerated for destroying — some would say repurposing — cottonwoods near Chelan. Sandy's handlers first matched her with a different blind date, a burly male named Hendrix, but the pair hadn't clicked. So Sandy was transferred into Chomper's enclosure, where she'd immediately moved in to his tin-roofed bungalow, nestled amongst his woodchips, and convinced him to share his apple slices. An arranged marriage, but sometimes those are the happiest.

Now, on a pale July morning six weeks after their betrothal, the two beavers were ticketed for a new home in Washington's wilds. Their human caretakers had coaxed them into separate cages for the journey. While Chomper shuffled in jumpy circles within his crate, his black nose twitching against the metal weave, Sandy hunkered down, snout pressed to floor, as though steeling herself against coming hardship. Catherine Means, a technician at the Washington Department of Fish and Wildlife, knelt to examine her charges. "She's been nervous since day one," Means said, sounding anxious herself. "I'm excited to get her back in the wild."

Sandy and Chomper were wards of the Methow Beaver Project, a collaboration between the state of Washington and the US Forest Service that, since 2008, has moved nearly four hundred rodents around the eastern slopes of the Cascades, the jagged range that bisects Washington on its way from British Columbia to Northern California. Most of the project's detainees fit a distinct profile: They've damaged private property — lopped down trees in an RV park, say, or inundated cow pastures — and landowners want them gone. Rather than calling lethal trappers, a growing contingent of citizens now notifies the project's crew, which captures beavers alive and relocates them to one corner of the Okanogan-Wenatchee National Forest, releasing the offenders into 1.3 million acres of glacier-hewn highlands dotted with lakes and patrolled by wolverines and cougars.

I'd come to the Methow Valley, a bucolic sweep of hayfields and orchards tucked in a sparsely populated corner of northern Washington, to witness the dawn of a new chapter in beaver-human relations. Skip Lisle and his Beaver Deceivers had taught North America to condone the presence of beavers; now the Methow team was demonstrating how to proactively insert the creatures into wounded landscapes, like field medics deployed to the front. "We want beavers up and down every stream, in all the head-waters," Kent Woodruff, the project's founder, told me as we crouched before his orange-toothed wards. In the way that some pet owners come to resemble their beloveds, I noticed that Woodruff, with his sturdy physique and whiskery muzzle, bore a passing resemblance to a beaver himself. He spoke in a low, raspy voice, firm with conviction. "We want these rivers dancing across their floodplains again."

Thanks in part to Woodruff's powers of persuasion, beavers haven't been in such high demand in Washington since fur hats went out of style. While other states continue to cast a suspicious eye upon the critters, in 2012 Washington's legislature passed a "Beaver Bill" touting their "significant role in maintaining the health of watersheds in the Pacific Northwest" and promoting "the live trapping and relocating of beavers . . . as a beneficial wildlife management practice."[1] In 2017 the state amended the law to encourage even more widespread transfers. Today more than half-a-dozen groups, agencies, and native tribes are transplanting beavers in the Evergreen State, the country's most sizzling hotbed of castor relocation. And nearly all of them employ techniques developed by Kent Woodruff and his merry band of Methowites, the tinkerers who've come closer than anyone else to mastering the occult art of convincing beavers to stay where you stick 'em.

Sandy and Chomper, then, had been abducted by the right crew. Catherine Means, with the help of her teammates, hoisted the two cages into the bed of a waiting truck, then draped them with a canvas tarp that reeked of glandular secretions from beavers past. The team piled the truckbed with freshly cut aspen branches — ready-made food once the pair reached their new environs — and pulled out of the hatchery, tires spitting pebbles as the pickup crawled onto Twin Lakes Road. The beavers were on the move.

———

Although the Methow Beaver Project may be the most ambitious group tackling beaver relocation in the United States, it's far from the first — nor is it the most colorful. For sheer ingenuity, no project, past or present, can match the Idaho Fish and Game Department's experiments with beaver paratroopers.

Idaho's efforts began shortly after World War II, when returning sailors and soldiers swelled the state's human population. Civilization encroached upon undeveloped terrain, particularly around McCall, a lakeside community in western Idaho. As new arrivals built homes and farms, they clashed with resident beavers, who destroyed orchards and stoppered irrigation systems. The Fish and Game Department, to its credit, resolved to relocate the rodents. But where would it put them — and how would it get them there?

Beavers, the state learned, resist easy transplanting. Fish and Game tried strapping cages atop horses, only to find the technique failed utterly. The beavers required constant cooling and watering; often the stressed animals refused to eat. The livestock weren't happy, either. "Horses and mules become spooky and quarrelsome when loaded with a struggling, odorous pair of live beavers," state biologist Elmo Heter reported, and who could blame them? "These problems involve further handling and too frequently result in a loss of beavers . . . a faster, cheaper, and safer method of transportation was a vital need."[2]

Fortunately, Heter, a resourceful fellow, knew that the recently concluded war had left behind a surplus of parachutes. He also had access to a Travel Air, a light-winged monoplane. Beavers made poor horseback riders; perhaps they'd prove better paratroopers.

Heter's plan was a stroke of quixotic genius: He proposed packing beavers into boxes, strapping them to parachutes, and air-dropping them over likely looking backcountry streams. He constructed his first crates of woven willow, a design that, he wrote, would allow an enclosed beaver to "gnaw his way to freedom" upon landing. The scientist abandoned that idea when he realized beavers were liable to chew themselves free while still inside the plane, and that a pack of beavers running loose within the confines of a small aircraft could only end in flames.

With willow out, the persistent Heter invented a different package — a suitcase-like crate whose elastic straps fell open upon impact. The onus of serving as crash-test pilot fell first to an old male beaver named Geronimo. Officers repeatedly pitched Geronimo's crate onto a landing field, trials that Heter relayed with a decidedly unscientific tone. "Each time he scrambled out of the box, someone was on hand to pick him up. Poor fellow! He finally became resigned, and as soon as we approached him, would crawl back into his box ready to go aloft again."[3]

All in all, the state air-dropped seventy-six beavers into Idaho's wilds in the fall of 1948, with extraordinary success. Only one beaver perished, an unfortunate casualty who somehow escaped in midair, climbed atop the plummeting crate, and fell to his death. When Heter surveyed the drop sites a year later, he found that "all the airborne transplantings [had been] successful. Beavers had built dams, constructed houses, stored up food, and were well on their way to producing colonies." (The scheme would strain credulity were it not for Sharon Clark, a historian who unearthed a grainy video in the bowels of the Fish and Game offices in 2015.) Altogether, each skydiver cost the agency $16 — around $160 today. "You may be sure," Heter added, "that 'Geronimo' had a priority reservation on the first ship into the hinterland, and that three young females went with him."[4]

Heter's zany triumph remains the most famous restoration effort in the history of beaverdom. There's a reason, however, that no modern-day scientists have replicated his beaver drop. Compared with contemporary castors, Geronimo and his fellow aerialists had a crucial advantage: The landscape into which they floated was nearly devoid of predators. By 1948 farmers, ranchers, and government agents had extirpated wolves in Idaho, and exiled grizzly bears and cougars to mountain redoubts far from the "small, open meadows" where Heter dropped his cargo. Conservationists eventually resuscitated large carnivores throughout North America; now, nearly two thousand wolves, two thousand grizzly bears, thirty thousand cougars, and millions of coyotes roam the West's hills and forests. Any paddle-tailed parachutist air-dropped today into the Northern Rockies would have to navigate a fearsome obstacle course of teeth and claws the moment she hit the ground.

Mark McKinstry, a latter-day Elmo Heter, knows the perils of beaver relocation better than anyone. In the 1990s McKinstry, then a biologist at the University of Wyoming, launched the most audacious relocation project since Geronimo plunged into the Idaho backcountry with three winsome damsels. McKinstry and his team jump-started their program by mailing surveys to more than five thousand Wyoming landowners. Where were beavers causing problems? Where had they been trapped out? Where might reintroducing them do good? "Before we started some big project, we wanted to make sure it would be accepted," McKinstry recalled to me. "It's kind of like reintroducing wolves in Yellowstone — you worry that nobody wants these animals."

As it turned out, many Wyomingites wanted beavers. Nearly half the landowners who responded to the survey, and all of the public land managers, had some interest in beaver reintroduction.[5] Wyoming's depopulated vastness also encouraged McKinstry's ambitions. "Wyoming has million-acre ranches where a landowner owns an entire watershed, from the top of a mountain to the intersection with a larger river," McKinstry said. "If beavers annihilate a few stands of cottonwood or aspen, those people don't care. You can work with someone like that."

Surveys complete, McKinstry began the arduous process of capturing beavers. From 1994 to 1999, he deployed snares and Hancock traps at thirty-three locations, setting his devices in the cool of early evening and checking them at dawn. McKinstry implanted around half his captives with radio transmitters, then trucked them to fourteen distant streams that ran through the property of tolerant landowners. Sometimes McKinstry's team hiked to remote headwaters with crates strapped to their packs; at more accessible sites, they backed up the trailer, flipped open the cages, and unceremoniously turned the animals loose. Altogether, 234 beavers graduated from McKinstry's program.[6]

By most measures, the endeavor succeeded. Colonies created wetlands in thirteen of the fourteen streams, tripling the average width of riparian areas — bountiful new habitats that attracted elk, deer, and so many moose that the state opened a special hunting season to prevent them from decimating willows. Duck surveys turned up seventy-five times more mallards, teals, widgeons, and gadwalls on beaver ponds than in beaverless areas.[7] Ranchers rejoiced at booming hay production and well-watered cows. Groups from the North American Wetlands Conservation Council to the National Rifle Association's conservation arm shook out their pockets to fund the project. McKinstry considers it "far and away" his career's finest achievement.

But while the campaign met its aims, its foot soldiers suffered discouraging casualty rates. Without a lodge in which to shelter, the kits and one-year-olds that McKinstry relocated were "almost DOA" — dead on arrival. Sensors in the beavers' tags alerted the scientists if they stopped moving, allowing McKinstry to track down the carcasses of his doomed rodents. Tracks, scat, hair, and bite marks told the unhappy tale. Around a third of McKinstry's beavers fell prey to black bears, coyotes, cougars, and other predators. "There was one spot, up by Cody, where it was like ringing the grizzly bear dinner bell," McKinstry told me ruefully. "A beaver is just a fat, slow, smelly package of meat."

Even the survivors, McKinstry found, tended to spurn the streams he'd chosen for them. More than half the radio-tagged beavers skedaddled for better habitat. Altogether, just 19 percent of the transplantees built dams in their new homes. McKinstry had to release an average of seventeen beavers per site to get one colony to stick.[8]

The lesson: Relocating beavers into carnivore-filled wildlands could work, but it wasn't particularly efficient — nor safe for the rodent refugees. Wyoming's transplanted beavers were both panacea and prey, miracle and morsel, savior and snack. There had to be a better way to keep the fat, slow, smelly meat packages alive.

———

Around the time that McKinstry was publishing his results, a biologist several hundred miles northwest had likewise begun contemplating the mechanics of beaver relocation.

In 2000, four years after Massachusetts banned most lethal traps, Washington State's voters passed a similar ballot initiative outlawing deadly body-grippers. With their lethal options limited, landowners began seeking more benign ways of handling beaver problems. In the Methow Valley, a two-thousand-square-mile watershed whose winding river drains to the mighty Columbia, they began contacting a US Forest Service biologist named John Rohrer.

They called upon a prepared mind. When Rohrer first came to the valley in 1991, he'd recognized that its dry, weedy meadows sorely missed the beaver dams that once moistened them. He enlisted brawny firefighters to construct ersatz check dams, to no avail. "Our dams worked well in summertime, backed the water up, and everything looked great," Rohrer told me. "But during fall and spring, you'd just get too much water coming through. They'd blow out and make the problem worse." The valley needed superior builders, but Rohrer feared that any introduced beavers would be wiped out by trappers — until the body-grip ban.

Rohrer learned to set nonlethal Hancocks from an old trapper, then dug a concrete pool in his backyard to hold his captives. The setup was dangerously ad hoc — one beaver chomped Rohrer's leg, imprinting an ugly bruise without breaking skin — but effective enough. (Rohrer has a proclivity for mucking about with dangerous wildlife: Soon after opening his personal beaver pond, he began capturing, harboring, and releasing terrariums full of nuisance rattlesnakes.) Rohrer loosed his beavers in the same weed-choked meadows where his own dams had failed. Within a year, dusty pastures transformed into lakes large enough to launch a canoe. Rohrer decided the Forest Service had to scale up.

The task of professionalizing Rohrer's backyard operation fell to his colleague Kent Woodruff. In some ways Woodruff was an odd choice to spearhead a beaver project: He had no particular experience with rodents, having spent his career studying bats, songbirds, and raptors. What he lacked in knowledge, however, he made up in desire. Although Woodruff had worked in the Methow since 1994 — excepting a stint in Washington, DC, where he helped the State Department coordinate humanitarian aid to Baghdad — he felt unfulfilled. During one project, he'd installed a network of nesting platforms for great gray owls and ferruginous hawks; although the structures attracted birds as intended, their designer wasn't satisfied. "Twenty nesting platforms just seemed like a tiny remnant of a legacy," Woodruff told me. He yearned to leave a more enduring mark upon the land. "When beaver came along as an opportunity to create a long-term improvement, I said, yeah, sign me up." He cobbled together a few grants, and the Methow Beaver Project was in business.

The problem, as in Wyoming, was predators. Cougars and black bears stalked the Methow, and wolves had begun moving in. Central Washington's conifer woods weren't quite as perilous as Wyoming's prairies — there were no grizzlies, for one thing — but the Methow Valley still presented a house of carnivorous horrors for a plump rodent. Beavers traveling on their own faced the gravest danger. Although Woodruff planned to trap and move complete family units, he knew that catching entire clans together wouldn't always be possible. Some beavers would inevitably be captured alone. It was those solo transplants who, once released, faced the gravest danger as they wandered upstream and down, lodgeless and in search of companionship.

Woodruff suspected that playing matchmaker could tackle those two concerns. If he could somehow pair up independent beavers, each duo, freed from the imperative of finding a mate, would be more likely to immediately settle down and construct a predator-proof compound. But how do you convince two strange beavers to fall for each other? What Kent Woodruff needed was a rodent love shack.

———

The morning before Sandy and Chomper were slated for release, I visited Woodruff's beaver motel for the first time. When the federal government opened the Winthrop National Fish Hatchery in 1940, it was the world's largest salmon factory; although it no longer holds that distinction, it might still be the most unusual. The hatchery churns out a million salmon and steelhead each year, and the facility's raceways — concrete-walled oval pens filled with running water — also make ideal beaver pens. In 2008 Woodruff commandeered a few unoccupied raceways and started moving in customers.

Woodruff's abiding goal was to make his wards' captivity comfortable. "This is the Hilton for beavers," he told me as we roamed the raceways, lined up side by side like bowling lanes. "No predators, good food, clean shavings." Beavers cruised their pens like submarines, bubble trails streaming from their matted fur, color-coded tags dangling from pinned-back ears. At the center of each raceway stood a little hut, a crude but cozy-looking cinder-block structure with a nest of wood shavings heaped on its floor. Black cloth stretched over triangular frames, like little pup tents, shaded beavers against the climbing sun. A battle-scarred male dubbed Half-Tail Dale squatted on the wooden ramp that led to his hut, eyeing us suspiciously. He held his dexterous hands curled against his chest and his damaged tail tucked beneath his haunches. (Most visiting beavers are named by friends of the project or visiting children. "There's usually a Justin Beaver," Catherine Means later told me wearily.)

Some inmates never find love — Half-Tail Dale, for one, had already been held for six weeks, and would soon have to be released lest his wild instincts atrophy. Male beavers tend to be farther-ranging and more rambunctious than females; as a consequence, the hatchery's prisoners tend to skew male. Some don't make it: Several beavers have died at the hatchery, struck down by infections they'd acquired in the wild, before their arrival. Still, most eventually meet their soulmate. "A lot of programs just catch 'em in one place and immediately release in another — there's no effort to move family units or form compatible groups," Michael Pollock, an aquatic ecosystems analyst at the National Oceanic and Atmospheric Administration who's among the Northwest's foremost beaver authorities, told me later. "The care with which they treat the animals is what's different about the Methow program."

The Methow team has also helped refine the half-art, half-science of beaver sexing. To set up Harry with Sally, of course, you have to know who's Harry and who's Sally. That's where beavers make matters difficult: Unlike, say, peacocks or anglerfish, beavers don't display *sexual dimorphism*, meaning that males and females are more or less indistinguishable in appearance. When mother beavers are nursing, their teats become visible; otherwise, not even the sharpest-eyed matchmaker can visually differentiate the sexes.

Examining beaver genitalia isn't revelatory, either. *Castor canadensis*, a misfit among mammals, lacks the external plug-and-socket genitalia shared by house mice, humpback whales, and human beings. Instead the oddballs possess modified *cloacas* — fleshy vents, analogous to the anatomy of birds and reptiles, that do triple duty in the departments of urine disposal, scent secretion, and reproduction. When beavers copulate, an act that takes place under the ice in late winter, the male's penis — which, like many mammals' members, contains a bone called the *baculum* — protrudes from the vent; the rest of the time, it's hidden from view. It was the beaver's invisible sex organs, incidentally, that earned it its scientific name: *Castor* comes from *castratum*, which, well, you can guess where that's from.

Despite beavers' inscrutable genitals, the Methow crew almost unerringly pairs up sexes. When I asked Woodruff how they pulled it off, he insisted roguishly that no visit would be complete without a hands-on sexing clinic. Forget visuals — my nose, he said, would know.

The tutorial would be conducted on none other than Half-Tail Dale, who, as though anticipating duress, hustled into his hut. Means rapped gently on Dale's tin roof to budge him, finally coaxing him into a cage. From there, the poor animal was nudged headfirst into a blue cloth sack, where he lay, bulky as a load of potatoes. Means peeled back the bag, unveiling Dale's hind legs, belly, and intricately pebbled tail, pale pink along the uneven, truncated edge where it had been mutilated, perhaps during some long-ago encounter with a boat's propellor.

"Okay, Ben," Woodruff deadpanned, his round face creased by stifled laughter. "You're up." I wrinkled my nose and prepared to sniff.

Unlikely though it sounds, consuming beavers' strong-smelling secretions is a tradition nearly as timeworn as wearing their furs. Castoreum has been used to tinge perfumes by scent-makers from Chanel to Shalimar; one contemporary fragrance encyclopedia describes its aroma as "wild and bodily, lustful and passionate, bestowing the one who wears it a delicate aura of sensuality."[9] Food companies, too, once included castoreum as an additive in yogurt, fruity drinks, candy, and other comestibles. Contrary to the claims of some food bloggers, there is virtually no chance that "beaver butt" tinges your vanilla ice cream today, though Swedish hunters can still avail themselves of bäverhojt, castoreum-flavored schnapps.[10]

For all their renown, however, castor sacs are merely beavers' most famous odor-producing organs. Beavers also possess *anal glands*, concealed, nipple-like lumps whose secretions guide the rodents through the world. Anal oil serves multiple functions — beavers deploy it to waterproof their pelage, grooming it into their coats using a special split toenail as a comb — but its primary role is communicative. Each beaver produces a distinct, fingerprint-like scent, which appears to encode nuanced genealogical information. In one 1997 experiment, biologist Lixing Sun exposed dispersing beavers to scent from two animals: one unrelated, the other a young sibling born after the disperser had already left her home colony.[11] The dispersers responded more territorially to the aroma of unrelated beavers – for instance, by pawing angrily at the scent mounds that Sun had marked – than they did around their siblings' secretions. Even though the dispersing beavers had never met their brothers and sisters, they recognized them as kin. Imagine trying to identify your separated-at-birth sister by sniffing her armpits.

We *Homo sapiens* are olfactory deadbeats by comparison, scarcely conscious of our noses when we're not walking past blooming hibiscus or fresh-baked brownies. Yet some of the secrets hidden in beaver anal glands can be deciphered by even our insensate noses. As Means held Half-Tail Dale's exposed nethers in place, another project biologist, Katie Weber, gave me a quick primer in anal aromas. If you smelled a hint of motor oil, she explained, you were holding a male. A whiff of old cheese indicated a female. Guys, she added, excreted darker, thicker fluid than gals. "Once you've done five, you can pretty well tell," Weber assured me.

Given that I had done exactly zero, I wasn't confident about my ability to determine Dale's sex – good thing he'd already been confirmed as male, and that this was only a practice run. I pressed my latex-gloved fingers against the damp fur of his belly as though feeling for tumors. Weber peered over my shoulder at the rodent writhing in my uncertain grasp. "Once you get the anal gland expressed, put some pressure on it, and you'll see oil," she offered, like a cornerman urging a boxer to lead with a left jab. The sack's darkness calmed Dale, though he occasionally kicked out with his clawed feet. I didn't blame him.

Beneath my fumbling fingers, the scent glands – angry twin volcanoes of pinkish flesh – finally popped out. A drop of amber liquid glistened on one tip. I dabbed at it gingerly with a tissue. Weber encouraged me to squeeze harder. I felt a bit like I was being hazed. "Be careful where you position yourself," called a biologist named Torre Stockard from behind the safety of a fence. "You're in the splash zone."

With a whispered apology to Dale, I pressed harder on his glands, which poured forth a viscous stream of caramel juice. ("Obviously this procedure should be done with your face a reasonable distance from the cloaca," one guidebook offers, "with your mouth shut."[12]) Weber swooped in to wipe up the mess. Dale's ordeal at my inexpert hands had mercifully concluded. "You're such a good model, Dale," Weber cooed.

She held up the tissue, blotched with hard-won secretion. Against my better judgment, I inhaled deeply. Motor oil? Maybe. But the bouquet also contained notes of overripe fruit, pet store interior, dead muskrat, paint, and countless other odors of murky provenance. It wasn't unpleasant, but it was powerful. Later, I learned that anal oil contains more than a hundred distinct chemical compounds, including many that are unique to males and females.[13]

Sniffing beaver butts may not sound particularly scientific, but it's effective. According to genetic analysis, Woodruff told me, the Methow Project has misidentified the sex of just one beaver since it began using the glandular technique, at Lixing Sun's suggestion, in 2011. Aided by the scrupulous sexing, the Methow project's relocation success rates have outstripped those of its predecessors. During his Wyoming experiments, Mark McKinstry burned through 234 animals to colonize thirteen sites; by contrast, the Methow's 360 beavers have built ponds in more than fifty separate locations. The Methow crew's ambitious site selection makes that number even more notable: They usually aim for higher reaches that the animals wouldn't readily infiltrate on their own. Timing matters as much as location. Release beavers in the spring, and they tend to scatter for better environs; transplant in fall, and they're more compelled to buckle down and build a lodge before the onset of winter.

In the years after the Methow Beaver Project's founding, its relocation techniques spread like wind-scattered seeds. The group hosted a series of beaver husbandry workshops, which became rites of passage for aspiring beaver traffickers — not only in Washington, but around the country, and even the world. Woodruff finally retired in 2017, ceding the program to Torre Stockard, a no-nonsense biologist with a background in emperor penguin research. At that winter's State of the Beaver conference, a gathering that draws the castorid world's leading lights to Canyonville, Oregon, every other year, Ben Dittbrenner, a University of Washington PhD student who's relocated dozens of beavers to the Skykomish watershed, devoted his first three slides to a Woodruff homage, Photoshopping his mentor's whiskery face onto the Godfather, Obi-Wan Kenobi, and the Most Interesting Man in the World. "The second we found out about Kent and contacted him, we were instantly part of this vast beaver network," Dittbrenner told me later. "He invited us to the Methow and went out of his way to ensure that we were making our project more efficient and successful. Without Kent, I don't know where we'd be."

———

No matter how skillfully you relocate beavers, though — no matter how accurately you sex them, how prudently you pair them, how carefully you select their new homes — the sad fact is that many, even most, are not going to stick. Dittbrenner is among the Northwest's most accomplished rodent movers, yet even he can't reliably get them to establish. "Some sites are phenomenal, and they're just going crazy," he told me. "Some sites looked like they were perfect, but we've introduced beavers there five to ten times, and they haven't taken."

Predation, the same problem that frustrated Mark McKinstry, still thwarts many relocations. At the State of the Beaver, I sat in on a talk by Vanessa Petro, a biologist at Oregon State University who outfitted thirty-eight nuisance beavers with tail-mounted radio transmitters and shuttled them around coastal Oregon in 2011 and 2012. Within four months more than half were dead, most victimized by mountain lions. One beaver clan bounced back and forth between hungry cougars so many times that Petro called it the "Ping-Pong colony." Altogether Petro's beavers built just nine dams, all of which eventually blew out.[14] "If our goal is to promote beaver populations and their damming activities," Petro told me later in an email, "shouldn't we be asking why the existing populations aren't doing it themselves?" Relocating beavers might seem like an easy way to colonize a watershed, but often there's a reason a seemingly suitable area has remained beaverless.

Such mixed results worry some beaver proponents, who fret that starry-eyed conservationists, in their justifiable enthusiasm for all things castorid, will reintroduce the rodents to streams that can't support dams, negating the very rationale of relocation. "We might put a lot of hope and capital and trust in the idea that a project is going to work, but if we don't have the right context we're setting ourselves up for failure in other places," Caroline Nash, a PhD candidate at Oregon State, told me. Nash is a member of a beaver study team whose emphasis is expectation management. In

2017 the group published a broad examination of beaver restoration on western rangelands that found, as she put it to me, that "practice is moving a lot faster than the pace of science." At least seventy-six separate projects are currently moving beavers around the West, yet only four monitor the results for longer than five years. Whether the rampant relocations are muddling genetics or transmitting diseases between colonies remains something of an unknown. And even well-meaning restorationists don't always select their release sites with the utmost care. Sometimes, the group found, organizations dump beavers in streams with inadequate food or unsuitable hydrology, consigning them to death or forced migration. Often, transplanted colonies simply vanish. Nash fears that failed relocations could lead to letdown — and ultimately throw cold water upon the greater cause of beaver-based restoration.[15]

Beavers, of course, are wild animals; to be wild is to be self-willed, to transcend the manipulation of human puppetmasters. Much though we may want beavers to settle where we put them and reliably filter our water and irrigate our crops, the creatures ultimately follow no designs but their own. Our relationship with beavers may have advanced light-years since the fur trade, but it's still fundamentally utilitarian; only now it's their dam-building skills, rather than their pelts, that we exploit. "We like beaver!" Nash insisted. "We just don't want the poor guys to get used."

Predators may stymie some projects, but cougars aren't the only threat to transplanted beavers: In many places trapping still makes relocation a nonstarter. Although the fur trade has rebounded somewhat thanks to demand in China, South Korea, and Russia, a recap of the 2017 North American Fur Auctions in *Trapping Today* reported that beaver pelts "averaged between $8–13, well below the cost of production."[16] While those basement prices have deterred some woodsmen, recreational trappers tend not to be driven by pecuniary motives. Trapping has deeper cultural and historical roots in this country than practically any other activity; to this day it's how many people commune with nature, the tradition they pass to their children. It's a form, a trapper told me once, of church.

As an avid fisherman, I have scant room to pass judgment on other folks' exploitative relationship with wild animals. I possess, too, nothing but respect for the extraordinary ecological knowledge tucked away in the memories of many trappers; I've met men whose functional knowledge of beaver behavior and biology equals any scientist's. And yet: When it comes to trapping management, all too often it seems like the left hand doesn't know what the right is doing. The sale of hunting, fishing, and trapping licenses generates up to 90 percent of the budgets for state fish and wildlife agencies;[17] when the need for revenue collides with beaver-based restoration, nonsensical policies can result. Drew Reed, the biologist who's transplanting beavers into Bridger-Teton National Forest to create swan habitat, is one of the guys caught in the middle. Even as the Forest Service trumpets beavers' benefits in the Bridger-Teton, the Wyoming Game and Fish Department allows trappers to take up to twenty-five beavers in the area.[18] Trappers sometimes call Reed to tell him they've captured an animal with his tags stapled in its ears.

"I have friends who kill-trap — I'm not against it," Reed told me. "But it would be great if we could meet in the middle and close a few areas so we can repopulate some of these drainages."

Idaho is another place whose relationship with beavers is replete with contradiction. In some parts of the state, the federal government pays ranchers to dig beaver-mimicking ponds for Columbia spotted frogs.[19] And the state's wildlife plan, completed in 2015, suggests beaver restoration could support forty "species of greatest conservation need," from steelhead to white-faced ibis to silver-haired bats, which snatch insects from the airspace over ponds.[20] "We call beavers a 'super strategy,'" Rita Dixon, the coordinator for the state's plan, told me.

Yet few areas remain entirely off-limits to trapping, and on many streams trappers face no limits whatsoever. Mike Settell, director of Watershed Guardians, told me existing regulations are scantily enforced. "If there's no enforcement and the trappers get wind of it — and they will — any relocations are just going to be a put and take," Settell warned when we met at his house in Pocatello, Idaho. On Settell's mantel, where other homes might display ceramic cherubs, I noted beaver figurines, chew sticks, and, in place of Jesus, St. Francis of Assisi, patron saint of animals. "I've actually argued against the idea of a beaver stocking program."

Settell's affection for beavers marks him as peculiar in Idaho, a state dominated by conservative farmers, ranchers, and hunters. Every winter Watershed Guardians conducts a volunteer-led beaver count in the Portneuf River watershed, a basin that drains twenty-one hundred square miles to

Pocatello's southeast. The most recent survey, Settell told me, found that colonies on many tributaries — Mink Creek, Jackson Creek, Birch Creek — had declined or vanished. Settell blamed poaching. "If you're out camping and you hear gunfire near the creek at night, I can almost guarantee what they're doing," Settell said. "If those people were poaching moose, they'd be doing jail time."

Washington's pro-beaver attitude notwithstanding, the rodents aren't much safer there. "There's still a lot of cultural fear," Joe Cannon, restoration ecologist at a Spokane-based nonprofit called the Lands Council, told me. "I hear the word *infestation* sometimes, like they're rats." The Lands Council is among the country's most prolific relocators: Since 2010 it's transplanted around 125 beavers in eastern Washington, establishing colonies at nine sites and attracting coverage from *The Atlantic, The Wall Street Journal*, and NPR. When I spoke to Cannon over the phone, though, he sounded downright dour about beavers' prospects. "There's all this hoopla, all this media, all these projects — but if anything, we're seeing a reduction in long-term beaver populations and dams and wetlands," he said.

The ballyhooed Beaver Bill legislation that Cannon himself helped spearhead may have enshrined the animals' value in state law, but it had done little to curb killing in the field. Between 2014 and 2016 the Washington Department of Fish and Wildlife recorded 7,698 beavers slain by recreational trappers and nuisance wildlife removers — over 2,500 dead beavers per year, an order of magnitude more than all the live relocations within the state combined.[21] "We're the only state to legalize the value of beavers, but we have greater protection on *tree squirrels*," Cannon grumbled. "If Washington is really leading the way, then we have a long way to go."

————

Once I'd narrowly passed my beaver-sexing test, Kent Woodruff decided I was worthy of experiencing the next stop in his Methow tour: an alpine drainage, high up dirt logging roads, called McFarland Creek. The team had released a pair of beavers there the previous spring; when we bushwhacked down to the stream, we saw the engineers had responded by erecting a dam half the length of a city block. Red osier dogwood sprouted from its ramparts. Jumbles of gnawed aspen clogged the new swamp, while a few doomed survivors stood helpless in the pond. A scum of pollen and mulch swirled lazily on the surface. Catherine Means and Katie Weber slogged into the backwater, sinking so deep into the sludge that the pond nearly overtopped their waders. They unfurled a tape measure and stretched it along the length of the dam. "Forty meters," Weber called out to Torre Stockard, who jotted the measurement on a clipboard. Weber turned to me. "This dam was twenty-four meters this spring. It's amazing how fast they do stuff."

"Everyone who likes to get a drink shows up here," Woodruff said. We caught the chatter of a Swainson's thrush, a warbling vireo, a song sparrow. Woodruff pointed out the pawprints of a black bear printed in the mud, and his face darkened. "I smell something dead, which is possibly not good," he said. He pivoted to Weber. "How much fresh sign are you seeing?"

"There's some new mud over on that side of the dam," Weber reassured him as she pulled herself out of the muck. "And some fresh cuttings over here. Looks pretty active."

Although wildlife habitat was McFarland Creek's most obvious virtue, another vital benefit was percolating beneath our feet. As the pond's weight pressed into the forest floor, the impounded water forced itself through the earth's cracks and pores, infiltrating soil and recharging the underlying aquifer. Some 30 percent of the world's fresh water is stored underground; in many rain-deprived places, aquifers — layers of soil and rock whose minute crevices are saturated with groundwater — provide the primary source. Our relationship with groundwater, as with most natural resources, tends toward abuse. Farmers above the Ogallala Aquifer, the formation that underlies much of the Midwest, have pumped their aquifer with such exuberance that in some spots only a few decades of water remain. In California's Central Valley, the nation's produce aisle, water users slurped up forty-one trillion gallons of groundwater between 1920 and 2013 — one-third the volume of Lake Erie.[22] In the 1970s, when the pace of pumping reached its furious apex, some farmland sank 30 feet as dewatered soil subsided, costing the state a billion dollars. Aquifers are often likened to savings accounts: Withdraw more than you make in interest, and pretty soon you're eating into your principal. Then you're broke.

To extend that metaphor, beaver ponds provide a fixed income, spreading water across landscapes and giving it time to penetrate soil. Beavers certainly can't recharge groundwater everywhere – some sections of the Ogallala, for instance, are characterized by such impenetrable rock that they can take millennia to recharge, leading scientists to dub their contents "fossil water." Where soil is sufficiently permeable, however, beavers could help replenish the same aquifers that we're depleting at breakneck pace. In the soggy fens of the Canadian Rockies, where beavers excavate wedges of peat to build long earthen walls, ponds raised groundwater levels by nearly half a foot over a wide area.[23] In Colorado dams pushed flows underground and kept downstream water tables elevated even during dry summer months.[24] And in eastern Washington researchers from the Lands Council estimated that beaver ponds held five to ten times more gallons beneath the surface than they did above it.[25]

For all that, we know surprisingly little about beaver-captured groundwater: The pond builders are incontrovertibly good for underground storage, yet precious few studies have attempted to slap a number on the benefits. What's clear, though, is that few places need beaver sponges more than central Washington. More than seven hundred glaciers freckle the slopes and coulees of the North Cascades, the most icebound area in the Lower 48. Those glaciers are supplemented by prodigious snowfall; in some years, the snowpack could bury a basketball hoop. The Cascades' glaciers store as much water as all of the state's rivers, lakes, and reservoirs combined. As temperatures climb each spring, the ice melts. Creeks swell into torrents, hurtling down mountainsides to meet ever-larger streams and rivers that eventually convey meltwater to farms and cities. Glaciers and snowpack continue to leak long after spring thaw, providing a time-release trickle that fills rivers well into fall, when high-country blizzards start the cycle afresh. All told, Washington glaciers release around 230 billion gallons of water each year, and snowpack contributes hundreds of billions more. If you have crunched into a Red Delicious apple, you have likely enjoyed a crop fed by Cascades runoff. And if you're a beer drinker, you certainly have: Around three-quarters of America's hops are raised in central Washington.

That makes "wet drought" all the more disturbing. When I came to the Methow in the summer of 2015, the region was still reeling from an eerie, balmy winter. Although plenty of precipitation had fallen, warm temperatures had produced rain instead of snow. Rather than recharging rivers throughout summer and fall, as gradually melting snowpack would, the aberrant winter rain had immediately raced downhill and vanished into the ocean. On April 1, 2015, snowpack in the Cascades stood at just 3 percent of the historic average, a record low. By summer rivers shrank to warm dribbles, irrigation districts shut down canals, and apple trees browned and brittled. For the Northwest's salmon runs, the season was nearly apocalyptic. In the Columbia River around a quarter-million sockeye salmon succumbed to warm waters and disease, many blotched with white fungus.[26]

Although snowpack in the Cascades rebounded the following winter, there can be no doubt about the direction of the trend. Washington's spring melt occurs earlier each year, average annual snowpack has declined, and more than fifty glaciers have vanished from Washington's peaks since the 1950s. The snow-starved winter of 2014–15 was less anomaly than climate change harbinger.

To Kent Woodruff and his Methowites, the wet drought was a call to arms. "Hundreds of billions of gallons come out of those snowpack faucets, but that reservoir is going dry," Woodruff told me, the ghost of a smile beneath his white stubble. "The question becomes, how can we capture and store that water? The propagandist in me says: I know."

———

The Methow crew designated Sandy and Chomper, the set-up lovers with whom this chapter began, for release into another remote drainage, this one called Bear Creek. Bear Creek had frustrated the crew before: The last beaver they'd released there had scrammed, tracing the Okanogan River nearly to the Canadian border, a two-hundred-mile picaresque. (The Methow's beavers have PIT tags, trackable microchips about the size of a grain of rice, implanted in their tails before release, allowing the crew to track their progress through the watershed.) "What's the motivation for a beaver to travel that distance? What was he looking for?" Woodruff mused as we rumbled uphill, as though trying to inhabit the *Umwelt* of his charges.

Catherine Means parked the truck in a meadow downstream from the release site. The last few weeks had been brutally hot, exacerbating the drought, and a blue haze from distant Canadian wildfires had settled over the basin like an unwanted houseguest. Grasshoppers hummed in the brittle brush. Woodruff warned Means not to park atop dry grass, lest the sizzling muffler spark a fire.

Wildfire is never far from the minds of the Methow's residents. In 2014 a blaze called the Carlton Complex, at the time the largest fire in Washington's history, devoured a quarter-million acres and destroyed 250 homes — Katie Weber's among them. Hellish though that inferno felt, it was only a prelude: Six weeks after my visit in 2015, the Okanogan Complex Fire howled through central Washington, charring another four hundred thousand acres. Three firefighters in nearby Twisp perished when the conflagration swallowed their fire engine, the same one on which all three of Woodruff's sons had trained. The climatic changes responsible for wet drought also stand to make bigger, hotter wildfires more common; in central Washington incinerated trees and gut-wrenching grief are rapidly becoming the new normal.

As we hiked up Bear Creek, the crew now toting Sandy and Chomper in their metal palanquins like royalty, we trudged through a torched landscape. Blackened lodgepoles clung to the steep walls of the draw; singed limbs littered the ground. As fire is wont to do, however, the Carlton Complex had generated new life as it destroyed old. In contrast with the scorched uplands, the vegetated creek corridor glowed a vivid green. A tangle of tender aspen suckers had sprung up in the fire's aftermath, freed from the shade of dead conifers. For all the destruction that fire had wrought, it had also catered a feast, an aspen bonanza that would allow beavers to recolonize this and other valleys across the Methow.

Not only would Sandy and Chomper enjoy the tasty growth, Woodruff said, they would also help fireproof the landscape by inundating meadows and drowning the flammable conifer understory. A wide, green, wet stream bottom would furnish an ideal firebreak. The idea of using beavers as fire extinguishers isn't a new one: During the Cold War, when the United States feared that the Soviet Union would drop incendiary explosives on the West, the military purportedly suggested repopulating beavers to foil the plan. The strategy never came to fruition, and the Soviets never dropped their firebombs.[27]

Higher and higher we marched, slapping aside thimbleberry and swatting mosquitoes, sliding down banks and splashing through shallows. Finally, in a shaded aspen glade, we reached the woody pièce de résistance, the secret sauce in the Methow's beaver-survival recipe: a ramshackle human-made lodge, custom-built for the new arrivals. Though the clumsy fort was a pale facsimile of the genuine article, it would buy the beavers time, protecting them from predators until they could construct their own home.

The team lowered the cages into a shin-deep eddy by the lodge's entrance. Sandy picked up her head and sniffed the summer air, revived by the tinkle of running water. "You don't know how it's gonna go until they get in there," Weber worried as she unlatched the cages. "When you come back and see them somewhere nearby, you get teary-eyed." The beavers waddled forth, vanishing into the lodge's dim recesses. Moments later a round of high-pitched beavery chortles floated up from the structure. Woodruff optimistically interpreted the sounds as squeals of delight.

The crew clambered out of the creek with dirt-stained cheeks, the empty cages swinging light in their hands. Woodruff may micromanage his beavers in captivity, but he seemed oddly dispassionate about their fate after deployment. "If we go back and see that there's an active pond, we don't care if it's two beavers or seven," he told me as we meandered back through the reborn aspen. Individual survival wasn't the point — the point was overall site establishment. Re-beavering drainages was less a goal in itself than a means to more anthropocentric ends: water storage, salmon recovery, fire mitigation. Beavers, in Kent Woodruff's hands, blur the line between wild animals and tools. They're willful, yes, but they're also doing our bidding. Yet I couldn't help but root for Sandy and Chomper not just as water-storing machines, but as individuals: slow, smelly meat packages trudging off into an uncertain future.

"The Methow Valley probably has 15 to 20 percent of its historic beaver population," Woodruff told me as we arrived at the truck. Returning to 100 percent, he knew, would never be possible — the human footprint was too vast, the opposition too fierce, the landscape too altered. Homes, roads, pastures and orchards jammed the valleys; streams that had once been as messy and debris-clogged as McFarland now followed the straight and narrow. Some rivers were beyond a beaver's help.

Yet even restoring beavers to around 40 percent of their former levels, Woodruff said, would fundamentally reshape the Methow Valley. Some watercourses were too far gone, yes, but others, particularly those high in the national forest, could yet be returned to marshy glory. How those streams flowed centuries ago, no one living could say; the beauty of contracting with beavers was that no conjecture was required. "We're not smart enough to know what a fully functional ecosystem looks like," Woodruff said, wiping his brow. "But beavers are."

— CHAPTER FIVE —

Realm of the Dammed

In early summer, as blizzards of cottonwood seeds drift through newly blue skies and Swainson's thrushes trill flutelike from the understory, salmon begin to surge up the rivers that enter Washington's Puget Sound. Into the Skagit, the Nisqually, and the Stillaguamish they flow: sea-bright coho, ruby-fleshed sockeye, titanic chinooks, humpbacked pinks, surfing the incoming tide past crumbling coastal bluffs and through seagrass deltas. Born in streams, salmon migrate to the sea as juveniles, grow fat on the Pacific Ocean's largesse, and return to their natal rivers to reproduce, coursing like blood cells up Washington's arteries and into the capillary network of spawning streams that vein the foothills of the Cascades and the Olympics. Their spent carcasses are rent by bears, pecked by eagles, and nibbled by mice; eventually they leach marine nutrients into the soil and fertilize towering conifers. In this corner of the world, Timothy Egan has written, "a river without salmon is a body without a soul."[1]

By that measure, the Snohomish River was once one of the Northwest's most soulful waterways. The Snohomish arises at the confluence of the Skykomish and the Snoqualmie, the mellifluously named rivers that drain the Cascades' western flanks, and wends twenty miles to Puget Sound, terminating in a braided salt marsh teeming with shorebirds and bivalves and fish. Native peoples thrived on the sound's bounty for thousands of years, harvesting crabs, clams, and, most important, salmon, whose annual arrival in early summer occasioned celebratory potlatch feasts. To the Snoqualmie, the Skykomish, and other Puget Sound tribes and bands, salmon were not merely resources – they were cultural partners, symbionts who loyally sustained their human dependents so long as the tribes protected their rivers and treated fish with due reverence.[2]

The intrusion of white settlers threatened to disrupt that ancient relationship. In 1853 President Franklin Pierce appointed Isaac Stevens governor of the new Washington Territory, and tasked the hardheaded Virginian with cajoling the Northwest's native peoples onto reservations to clear the land for colonists. On January 22, 1855, Stevens and tribal chiefs signed the Treaty of Point Elliott, an agreement that forced the Puget Sound's native tribes onto the 22,000-acre Tulalip Reservation. Although one judge later called Stevens's treaties "unfair, unjust, ungenerous, and illegal,"[3] the deals did have a redeeming feature: They permanently preserved tribal members' rights to fish at their "usual and accustomed" places, including sites on state, federal, and private land. But the treaties were seldom respected: When natives sought to exercise their rights in the 1960s and '70s, they were arrested by Washington State officials and harassed, sometimes violently, by their white counterparts. (One common refrain: "Save a Salmon, Can an Indian.") In 1974 a federal court finally intervened in the Fish Wars on the tribes' behalf, granting the Northwest's native peoples half the catch and authority to co-manage their fish with the state of Washington – a ruling known, for the judge who authored it, as the Boldt Decision. It led, decades later and in a roundabout way, to beavers.

Terry Williams, a genial, gravel-voiced tribal member who wears a dark mustache and his hair pulled tight in a ponytail, grew up on the Tulalip Reservation in the 1960s. Even then, Williams had a passion for aquatic rodents: He and a young cousin caught juvenile beavers alive, tracking them with dogs and wrangling them into sacks, and relocated them to wetlands near their home – for no other reason, he told me, than that he was "thirteen years old and curious." They caught raccoons, too, which they kept in the house. "My mom's sister went a little crazy when they started going through the cupboards."

After a stint in Vietnam, Williams returned to Washington to work for a railroad company and attend college via the GI Bill. He tried commercial fishing in Puget Sound for a season, netting salmon to sell and flounder to take home. In the early 1980s some friends asked if he'd consider working for the tribe's fisheries department – just for a year, they promised. The

Tulalip Tribes were then embroiled in legal struggles all over Washington State in defense of its members' fishing rights, and Williams spent seven days a week on the road, preparing arguments and sitting in on court sessions. A year on staff turned into two, ten, a career. Today Williams serves as the tribe's treaty rights commissioner, and has, at one time or another, held a seat on every board, council, and commission pertaining to salmon recovery in Puget Sound, and many beyond. "I just got addicted to it," he told me.

It's easy to understand why Williams found salmon management intoxicating. It was an infinitely complex struggle, equal parts legal, social, political, and ecological; at stake was nothing less than his people's economic prospects and cultural survival. At its outset, the Fish Wars were waged over a fairly straightforward conflict: The tribes wanted access to their physical fishing sites and a guaranteed portion of the catch. Once those rights were secured, at least on paper, the battleground shifted to fish recovery. The Puget Sound's salmon had collapsed, the victims of decades of dams, overfishing, and development; Williams grew up eating not salmon but government-issued rations, including bug-infested flour and butter dyed with yellow food coloring. Thousands of acres of marsh had been paved over, hundreds of embayments wiped out. Beaches had been bulwarked, lowland forests demolished.[4] What use was having your right to fish confirmed by the courts if there were no fish to catch?

Around 2007 Williams began to think, for the first time in decades, about beavers. He recalled his childhood experiments in beaver relocation, and how ponds expanded when he and his cousin installed their rodent captives in wetlands on the reservation. Williams had recently become worried about a new threat to Puget Sound's salmon: climate change. He began to wonder if beavers, by capturing water and creating ponded shelter for juvenile salmon, could sustain the sound's fish.

One typically misty northwestern morning, I drove to a nondescript Seattle suburb before dawn to observe the fulfillment of Terry Williams's vision. I met Molly Alves and David Bailey — biologists in the Tulalip Tribes' natural resources department, though not tribal members themselves — at a gray, duck-dotted wetland, where a flood-fearing landowner had reported a beaver colony. "We'll get a lot of calls from homeowners' associations saying someone saw a beaver again," Alves told me as we trudged down to the water's edge. "I'll say, So, what's the issue? 'Well, we *saw* it.'" She rolled her eyes. "Seeing an animal does not constitute a nuisance. More often than not, we try to convince people to let beavers stay."

In Seattle's fast-growing King and Snohomish Counties, however, where lots of people live near lots of water, some conflicts could only be solved via relocation. On this morning the beavers had eluded capture, although the crew's Hancock trap had slammed shut overnight. Alves, a lanky, affable scientist, sniffed the damp air. "It just reeks of beaver," she said with a frown. She suspected a near miss — maybe a tiny kit that had slipped free from the Hancock. Alves wiped her hand on the trap's frame and held it out to Bailey.

"Smell it," she said with a grin.

He recoiled. "I'm not gonna smell your hand."

"See if you can tell what sex it is."

Bailey leaned in and wrinkled his nose. "Female?"

Alves tutted disapprovingly. "Male."

Bailey sighed. "Do you want some hand sanitizer?"

For all the fun Alves and Bailey seemed to have on the job, it was serious work. Since 2013 Tulalip biologists, leaning on protocols adapted from the Methow Valley, have transplanted more than a hundred beavers from Puget Sound's densely developed lowlands to tribal treaty lands in the Mount Baker–Snoqualmie National Forest, a wonderland of cedars and Douglas firs that straddles the North Cascades. The forest's steep slopes and crosshatching of logging roads prevent beavers from reaching the forest's headwaters on their own, making the dense woods a perfect candidate for release.

After our fruitless trapping session, Alves and Bailey drove me along the Skykomish River, roaring and turquoise in its narrow timbered valley, to their favorite release site, a nameless tributary ringed by bigleaf maple. A colony of relocated Tulalip beavers had conjoined several separate streams into a mirrored pond, nearly the size of a baseball field, from which side channels radiated like spokes from a hub. "You know it's succeeded when you need a flotation device to monitor your site," Bailey half joked.

The beavers' arrival had been a boon for wildlife: Motion-activated cameras had caught bobcats, coyotes, otters, bears, and mink slinking through the new wetland, many tiptoeing along the crests of dams as though traversing a balance beam. But the most dramatic changes occurred below the surface. Before the Tulalip Tribes restored beavers to this place, the skimpy wetland had stood fishless for years, largely cut off from the nearby stream. As beavers broadened the pond, though, it rejoined the creek, swelling from an isolated pocket of water to a connected side channel. And the fish had followed: Within two years of the rodents' arrival, Alves spotted fry sheltering in the pond.

Even so, Alves and Bailey were little prepared for the dozens, maybe hundreds, of juvenile fish that teemed in the pond on the day of our visit. The multitudinous babies darted into nooks and crevices within the dam at our approach, white flashes along their anal fins betraying them as young coho. Beavers, I realized, had reunited this tiny pond high in the western Cascades to a vast and interdependent coastal ecosystem. Some fraction of these fry would survive to reach Puget Sound, where they'd grow fat on krill and anchovies, enter the nets of tribal fishermen, star in potlatches, and transmit a durable culture across generations.

"It's kind of blowing my mind how many fish are here," Alves said, her voice full of wonder. "From the Tribe's perspective, this is what it's all about."

To their credit, the Tulalip Tribes have not merely been content to reintroduce beavers themselves — they have also expended considerable political capital to help advance the larger cause of rodent restoration. In 2012, recall, Washington passed its "Beaver Bill," the law that permitted relocations and turned the state into a nexus of castorid activity. Though the legislation was well intended, it contained a flaw: While it permitted biologists to move beavers to sparsely settled eastern Washington, it prohibited releasing them west of the Cascades — the region that's home, of course, to Seattle, Tacoma, Olympia, and the state's other population centers. The message: Beavers can do good, but keep the damn things away from people.

Thanks to their salmon rights, the Tulalip weren't bound by the prohibition: Several years earlier, the tribes had struck a deal with the federal government to co-manage watersheds within the Mount Baker–Snoqualmie National Forest, granting the tribe authority to restore habitat as it saw fit, beaver relocations included. Still, the law's illogic irked Terry Williams. More beavers on more Washington rivers would mean more Puget Sound salmon, he figured. In 2017 the tribe dispatched a lobbyist to Olympia to advocate for a revision to the Beaver Bill. The stubborn lawmaker who'd pushed for the ban on western Washington relocations had retired, and the revised bill sailed through. No longer were the Tulalip Tribes the sole entities capable of moving beavers around western Washington — the doors had been flung open.

Williams couldn't have been more pleased at the outcome, though he wasn't surprised. When he began his career in fisheries management, he told me, he'd found himself bound by a skein of laws intended to thwart tribal fishing. Decades ago he'd griped to the tribe's chairman about the bevy of legal obstacles. "Well, that's not so difficult," the chairman retorted. "If the law doesn't work for you, change it."

Williams chuckled hoarsely as he recounted the story. "Because of that simple statement," he told me, "I've changed so many laws I can't count 'em anymore."

———

To Terry Williams, the fact that beavers create salmon habitat is so self-evident that it's worth changing the law. In other quarters, however, the relationship between the mammal and the fish remains a point of contention. Throughout the Pacific Northwest, salmon are the primary rationale for beaver restoration, yet the intransigence of some fish biologists still impedes the rodent's return.

The origins of that skepticism are hard to untangle, but it probably has much to do with wood. For modern paddlers and fly fishermen accustomed to free-flowing rivers and streams, it is impossible to imagine the woody wrack that once cluttered American waterways. Trappers and explorers found many watercourses blockaded by impenetrable logjams composed of gargantuan old-growth trees. Puget Sound's Skagit, for one, less resembled a river than a lumberyard. "Tier upon tier of logs up to eight feet in diameter, and packed solidly enough to be crossed almost anywhere, formed a stable obstacle that supported a forest of 2-to-3-foot-diameter trees growing on its surface," wrote David Montgomery in his book *King of Fish*.[5] All that woody debris, Montgomery added, created prime fish habitat: "Perennially submerged wetlands and sloughs provided ideal summer rearing habitat and slow-water refuges for salmon during winter floods." Beavers don't get credit for the logjam: Not even the most inexhaustible chewer could take down the massive firs and cedars that formed its superstructure. But beavers abounded in the Skagit watershed, and their upstream gnawing surely contributed to the jam's mass. (In Quebec, Bob Naiman found that beavers mobilized more than half the willow and aspen that clogged streams.[6]) And the 150 square miles of Skagit Valley that the epic logjam flooded must have been a glorious beaver-and-salmon playground.

But American rivers did not remain so thoroughly jammed. In the late 1800s the US Army Corps of Engineers, fixated on turning rivers into freeways for shipping, embarked upon an anti-logjam crusade. On the Skagit, Stillaguamish, and Snohomish Rivers, corps "snagboats" extracted more than 150,000 logs.[7] As industrial logging intensified in the twentieth century, the war on wood shifted battlegrounds. Profit-minded loggers dumped unmarketable lumber into rivers, where the woody waste created unsightly logjams and scoured holes. At first state agencies cleared streams only of logging detritus, which they feared impeded the upstream passage of spawning salmon. Soon, however, that well-intentioned practice transformed into an all-out campaign against in-stream wood, no matter its provenance. In 1972 Oregon passed a law mandating wood removal, and Washington and California followed suit. "All along the West Coast," wrote Montgomery, "a clean stream not only looked like a good idea, it was the law."[8] Beaver dams, in many cases the most visible blockages, were not spared.

Biologists eventually realized the folly of extracting wood for salmon's sake. But the fallacy that beavers and fish couldn't coexist persists – not only in the Northwest, but everywhere that salmonids thrive. The Miramichi Salmon Association, devoted to restoring Atlantic salmon in New Brunswick, cuts notches in dozens of beaver dams each year to assist returning spawners upstream.[9] The Forest Service has been known to destroy dams on Lake Tahoe tributaries to clear the way for stocked kokanee salmon – demolishing a native mammal's works to advantage an introduced fish.[10] And even that nuttiness pales in comparison with a 2009 proposal funded by the Atlantic Salmon Conservation Foundation, which suggested trappers eradicate beavers from ten river systems on Canada's Prince Edward Island and enforce "beaver-free zones" in others.[11] The report, based heavily on anecdote and conjecture, was never acted upon in Canada, but it influenced policy across the Atlantic: Scottish sportfishing groups referenced it to oppose beaver reintroduction in Britain.[12]

The fish fervor reached its apex in northern Wisconsin, where, from 1993 to 2014, Wildlife Services eliminated more than sixteen thousand beavers and dynamited thousands of dams to "rehabilitate" habitat for brook trout — an Orwellian policy that makes one wonder how the poor brookies survived before benevolent trappers came to their rescue.[13] The Badger State's castor-killing campaign has been guided primarily by a 2002 study suggesting that controlling the rodents, and converting their pond complexes into free-flowing streams, helped trout grow larger and more abundant.[14] Critics counter that the study lacked comparison streams, that its statistical analyses suffered from fatal flaws, and that trout populations swelled for different reasons, like cleaner water or stocking.[15] Other Wisconsin-based research has found that, contra the state's fear that beavers heat up creeks by felling shade trees and exposing ponds to sunlight, dams little affect stream temperatures.[16] "To keep every cog and wheel is the first precaution of intelligent tinkering," wrote Aldo Leopold; one wonders what Wisconsin's most beloved ecologist would have made of his state's hyper-aggressive approach to beaver management.

In fairness, beaver dams *can* pose a temporary obstacle to migrating fish, especially when flows drop in the fall. Usually, though, fish pass the blockades without much trouble. A Utah study that tagged over thirteen hundred trout (some of which the scientists caught on hook and line, proving that research doesn't have to be tedious) found that native cutthroat easily negotiated even large dams. Non-native brown trout, meanwhile, had more trouble — suggesting that beavers could be a valuable tool for preserving a stream's indigenous fauna.[17] Fish have plenty of clever methods for circumnavigating beaver works. They often bypass dams via side channels, like motorists avoiding highway traffic by taking local roads. Sometimes they wait patiently in the plunge pools below dams for high flows. Adult salmon may simply soar over barriers; the Atlantic salmon, *Salmo salar*, isn't dubbed "The Leaper" for nothing. Rebekah Levine has observed adult grayling — trout cousins with colorful sails for dorsal fins — squirming through Montana dams like children navigating a jungle gym. "They just wriggle right through," she told me, still amazed.

Befitting their reputation as a keystone species, the munificent rodents actually help fish in many ways. Beavers mitigate drought: When western Wyoming dried up in the early 2000s, researchers found that young cutthroat trout survived best in rodent-created pools in a place called, fittingly, Water Canyon.[18] Beavers make fish food: Bob Naiman found that ponds contain up to five times more invertebrates than open channels, an almost unfathomable seventy-three thousand bugs per square meter.[19] And while fish folks sometimes complain that silty pond floors make lousy breeding habitat for salmon and trout, which prefer rocky bottoms, every particle that gets trapped by a beaver is a particle that won't smother spawning gravel downstream. During 2001 floods, three beaver dams in Russia trapped 4,250 *tons* of solids — about twenty blue whales' worth.[20]

It gets better. Beavers — perhaps ones with delusions of becoming sea otters — build dams in the Skagit River estuary, a brackish marsh inundated twice a day by Pacific tides. When the sea comes up, the dams vanish; when it goes out, the structures reemerge, trapping the ocean and allowing their creators to navigate underwater even at low tide — and providing prime shelter for young chinook salmon and other fish.[21] Elsewhere beavers carve out salmon habitat without even building dams. In 2014 and 2015 Marisa Parish snorkeled among beavers' works on California's Smith River for her PhD dissertation at Humboldt State University, peering into the dark entrances of submerged bank burrows with a flashlight. "Sometimes you can almost get your whole body into the entrance of the burrow, and your heart gets racing pretty fast," she recalled to me. Although she may not have been comfortable in the burrows, fish certainly were: Parish found four species of juvenile salmon taking cover in the underwater enclaves.[22]

In 2012 a group of British researchers, led by Paul Kemp, waded through the morass of scientific literature to settle the matter once and for all. Kemp reviewed 108 published papers, discovering that scientists cited beaver benefits to fish much more frequently than they did negative consequences. What's more, while the majority of beaver benefits — improved habitat complexity, steadier flows, jacked-up insect production — were grounded in hard data, more than 70 percent of the purported detriments were merely speculative. Decades of grist from the anti-beaver rumor mill had congealed into unchallenged truisms. Even scientists, it turns out, can be awfully unscientific.[23]

I'll cop to oversimplifying a complex issue here, but at an intuitive level, it's ludicrous to me that the harmony of beavers and fish remains up for debate. Before fur traders and colonists trashed the place, North American streams were stacked with hundreds of millions more beaver dams than they are today — and yet our rivers so churned with fish that European colonists claimed they "ran silver." Castorids and salmonids — along with other fish families, of course — possess millions of years of entwined evolutionary history. Just as Nick Haddad's butterflies evolved to exploit beaver meadows, so must fish have adapted to the dams, ponds, and wetlands that shaped their habitat. The evolutionary connection is so commonsense that I've seen it boiled down to a bumper sticker: BEAVERS TAUGHT SALMON TO JUMP.

If beavers promote fish production, it stands to reason that the rodent's near-extinction would have been catastrophic for the Pacific Northwest's salmon — for the region's soul itself. Fish biologists often refer ruefully to the "Four H's," categories of human-inflicted harm that nearly obliterated many salmon runs. *Harvest*, or overfishing, converted salmon flesh to cold cash. *Hydroelectric dams* blocked upstream migration, sacrificing salmon for cheap electrons. *Hatcheries*, fish factories built to rehabilitate flagging populations by churning out more babies, inadvertently overwhelmed some wild runs with genetically inferior stock. All the while, salmon suffered from the most insidious H of all: *habitat loss*, the incremental destruction of tributaries in which salmon breed, hatch, and grow up. Mining contaminated mountain creeks with toxic tailings; logging destabilized hillsides and smothered spawning gravels; irrigation diversions slurped streams dry. The Northwest's salmon habitat was slain by a thousand cuts as one stream-mile at a time eroded, unraveled, or vanished altogether.

We tend to think of habitat loss as a byproduct of permanent European settlement, a disaster set in motion by farms and cows and mines and roads. In truth the Northwest's salmon streams began collapsing well before the first intrepid homesteaders rolled into Oregon in the 1840s. The deterioration of salmon habitat began decades earlier, with a series of ambitious trapping sorties that de-beavered much of the inland Northwest. These fateful expeditions were driven not merely by short-term profits but also by geopolitics — a cold war, fought over furs, whose outcome helped shape the American West.

The clash was set in motion in 1818, the year that the fledgling United States signed a treaty with Britain pledging to share the Oregon Territory — a vast swath of country, little explored by white people, that included modern-day Oregon, Washington, British Columbia, Idaho, and parts of Wyoming and Montana. The two nations had long sparred for control of the Northwest, a tiresome dispute centered on who'd planted which flags where and when.

(That the region was the homeland of the Umatilla, the Nez Perce, the Shoshone-Bannock, and dozens of other tribes didn't seem to much concern anyone.) The 1818 bargain kicked the can down the road by instating "joint occupation," during which both British and American citizens could live and trade in the Oregon Territory. But the treaty failed to resolve the most basic question: Where should the border between the United States and Britain eventually be drawn? The Americans wanted the border to follow the 49th parallel, the straight line that today divides the United States from Canada to the Pacific Ocean. The British, meanwhile, advocated for setting the border along the Columbia River — an arrangement that would have raised the Union Jack over much of present-day Washington State.[24]

To British traders, the joint occupation represented both threat and opportunity. Canada's fur industry was then dominated by the Hudson's Bay Company — a beaver-trading behemoth that had commanded the continent for so long that its initials, *HBC*, were said to stand for "Here Before Christ." The company's governor, George Simpson, coveted the contested lands around the Columbia, but he knew that HBC's grip was fragile. American trappers would soon cross the Rocky Mountains and penetrate Oregon, and farmers and settlers would not be far behind. Once American homesteaders had flooded the country, Simpson feared the border would be drawn in their favor — and to the company's detriment.

The solution that Simpson hit upon was shrewd and merciless. If the beaver threatened to lure American trappers to Oregon, the Hudson's Bay Company would simply destroy the beaver. Simpson proposed to enact a "fur desert" south of the Columbia River, a barren buffer of beaverless country whose destitution would dissuade the American hordes from proceeding farther north.[25]

Simpson's scorched-earth proposal flew in the face of the Hudson's Bay Company's usual approach to stewardship. Its agents preferred to "nurse the land": Elsewhere the company imposed strict beaver quotas to prevent overharvesting, restricted hunting seasons, and sold native people discounted provisions so they wouldn't kill beavers for food.[26] Rather than nursing the so-called Snake Country south of the Columbia, though, Simpson vowed to "denude" it. "If properly managed," he wrote to headquarters in London, "no question exists that it would yield handsome profits as we have convincing proof that the country is a rich preserve of

Beaver and which for political reasons we should endeavor to destroy as fast as possible."[27] And he knew just the man for the job.

Fur trappers may have been, as Don Berry put it, "a majority of scoundrels," but Peter Skene Ogden was more scoundrellish than most. Born to affluence in Montreal in 1794, Ogden, a man of "less than average height and greater than average width," forsook a life of comfort for the thrill of the fur trade. He quickly distinguished himself through his cruelty and cunning: He thumped rivals with sticks and fists, brandished guns at every opportunity, and, in 1816, murdered an Indian at a trading post. Other traders considered him "an irritating, irresponsible individual who was capable of the most savage and cold-blooded crimes."[28] Yet Simpson, a brutal man in his own right, was impressed by his colleague's grim acumen; when he wrote that Ogden's "conduct and actions are not influenced or governed by any good or honourable principle," he meant it as a compliment. Who better to lead a campaign of wanton destruction than "the most unprincipled man in Indian Country"?[29]

Ogden did his job well. During six expeditions from 1824 to 1830, the stout woodsman led a ragtag band of Canadian voyageurs and mixed-race trappers as far east as Utah (where rivers, valleys, and cities still bear his name) and as far south as the Colorado River. Ogden's explorations literally put much of the West on the map, clearing up the twisted geography of river systems like the Willamette, the Umpqua, and the Bear and paving the way for future explorations.[30] He was the first white man to document the Humboldt River and Mount Shasta (both of which, of course, had been known to indigenous peoples for millennia), and in 1829 he made the first formal reconnaissance of the eastern Sierra Nevada.

But destroying beavers, not making maps, was the primary raison d'être of Ogden's journeys. While Simpson's fur desert policy held, the company hauled thirty-five thousand pelts from the Snake Country — doubly astounding, given the aridity of eastern Oregon and Idaho.[31] Ogden occasionally seemed sorrowful about his role as exterminator. "It is scarcely credible what a destruction of beaver by trapping at this season," he wrote in May 1829, "within the last few days upwards of fifty females have been taken and on average each with four young ready to litter."[32] Elsewhere in the company's territory, the wanton killing of pregnant females would have been forbidden. But Ogden's qualms did not prevent him from executing Simpson's ruthless tactic. By the time his successor, John Work, traveled to the Snake Country in 1830 and 1831, the land was completely denuded; Work's party went more than five months without recording a pelt.[33]

For his part, Simpson seemed elated. "It is highly gratifying to be enabled to say that all opposition from citizens of the United States is now at an end," he wrote.[34] It's true that the Hudson's Bay Company won the battle: The Snake Country expeditions succeeded in staving off American trappers for a time, and made Simpson and his associates a pile of money. But the Yanks won the war. In 1846 the border was fixed at the 49th parallel, the boundary the United States had sought.

Simpson's fur desert was a vile strategy, but the Americans were scarcely more enlightened. During Peter Skene Ogden's travels, he often came upon streams that had been plundered by American parties. "He may have wondered if the Americans were beating Hudson's Bay at its own policy and denuding the land themselves," wrote the legal scholar John Phillip Reid.[35] If the Hudson's Bay Company had nursed the land, as per their normal approach, perhaps the region would have retained some beavers; or maybe the Americans simply would have killed what the Canadians spared.

Either way, the fur desert's ecological legacy lingered. As the historian Jennifer Ott wrote:

> Over time . . . a lowered water table, lost surface water, and increased erosion resulted from the reduced number of beaver colonies. The consequences went beyond the vegetation to the animals that relied on riparian areas for water, food, and habitat. Figuring an average of six beaver per colony, the thirty-five thousand taken out by the HBC represents the equivalent of nearly six thousand beaver ponds. . . . [T]he water ran differently, large mammal forage was sustained differently, water tables were maintained differently, and creatures that lived in beaver ponds had to find new bodies of water to call home.

And few creatures were more harmed by the beaver's eviction than salmon.

———

Among the waterways that Ogden's band plundered was the John Day River, a 280-mile-long Columbia tributary named after a fur trapper primarily famous for going insane while seeking a route through the Blue

Mountains. Despite losing its rodent architects, the John Day today hosts a substantial run of endangered steelhead, chrome-sided rainbow trout that, like salmon, pass their adulthoods at sea and return to their rivers of origin to spawn. In 2015 a few hundred returning steelies evaded fishermen and sea lions at the Columbia River's mouth, navigated ladders through three giant dams, forked into the John Day, and, by fall, found their way into a slender stream called Bridge Creek. There they courted, excavated gravel nests, and deposited globular orange eggs no wider than the head of a nail. The eggs hatched into tiny *alevins* still joined at their belly to nourishing yolk sacs; grew over the next year into glittery fry; and, in the summer of 2017, were rudely abducted by a biologist named Jake Wirtz.

I joined Wirtz and his crew on the banks of Bridge Creek one June morning. Two hundred miles west, the Columbia greets the Pacific Ocean in a sweeping gray estuary, the Oregon coast brooding and gauzy. Here in central Oregon, though, the river's tributaries zigzag through the rain shadow cast by the Cascade Range and the land shrivels to high desert, all etiolated sagebrush and black juniper. An evening thunderstorm had tamped down the dust, and the morning dawned cool and clear, redolent of crushed sage and cut hay. A seashell-pink sunrise chased night over the hills.

The team had no time for ogling scenery. The day, already, was wasting. Waders on, nets and buckets in hand, the quartet — Wirtz, Sam Simmons, Austin Decuir, and Devin Baumer — stumbled down the eroded bank and into the snarled, watery chaos of Bridge Creek.

Within the stream corridor, a jungle of willow blotted out the dawn. Rather than a single channel, Bridge Creek boasted many. Twisting braids of rushing water converged, tangled, and pulled apart like strands of unbrushed hair. Walls of head-high grass ushered us back and forth through waist-deep side channels, a watery hedge maze. A coiled rattlesnake bzzzzed from the brush. After pausing to take notes, I looked up to find my companions had vanished in the wet thickets. I heard Wirtz's voice, faint over the stream's surround-sound gurgling: "Ooh, I can feel that spider right on the back of my neck."

The crew had entered this ravine, braving snakes and arachnids, to do some electrofishing, one of biology's oddest techniques. Wirtz, stocky and bearded, charged down riffles and staggered through pools like the collegiate fullback he'd once been, a bulky battery strapped to his back and a metal probe hefted in one hand. The assemblage generated an electrical current, which stunned any fish caught in its field; the fish, once shocked, floated helplessly downstream, where the crew intercepted them with a seine net stretched between wooden poles like a scroll. The brief electrocution probably wasn't pleasant, but it didn't injure the fish. As soon as they were released, they slipped like mercury into the latte-colored stream.

The morning started slow, the probe turning up only dace and suckers. Soon, though, Wirtz came to a deep, promising hole, its surface churning like a pot set to a rolling boil. The pool, I saw, had been created by a derelict beaver dam, which deflected the creek's flow to starboard and scoured a pocket from the muddy bank. Examining the structure, I noticed that it was buttressed by vertical posts a bit too smooth and straight to have been carved with incisors. Beavers had worked on this dam, but they'd evidently collaborated with people.

Wirtz plunged into the hole, waving his probe. The crew hoisted the net, and sure enough, a finger-length steelhead flopped in the mesh. It was a lovely fish, its silver back spangled with black freckles and dark parr marks, its flank rosy from tailstock to gill cover. "That's what we're looking for right there," Wirtz crowed. He glanced upstream, seeking more enticing spots. I noticed another strange hybrid dam in a state of semi-repair, water gushing through its posts. "Right now, early morning, they're sitting on the seam looking for food, hanging out in slow water and surveying what's floating past them," he said. "They need access to microhabitat, to different stream features: slow water, fast water, woody cover, riffles, pools." Into the bucket went the steelhead, the day's first capture.

The electrofishing picked up, steelhead after steelhead flashing in the net — "Hey, trifecta!" Wirtz hollered after one three-fish haul — until the bucket housed a squirming school. The crew adjourned on a gravel bar. Baumer plucked steelhead from the bucket, handling the fish as adroitly as a blackjack dealer manipulates cards. She dipped them briefly in clove oil to calm them, measured their lengths and weights, injected their bellies with an identifying microchip, and dropped them into another bucket. From there they'd be released, their movements now tracked by an array of electronic tag readers mounted throughout the watershed.

Baumer, a five-year veteran of the project, had the routine down so pat she could joke as she worked. "What's a pirate's favorite letter?" she asked.

"Arrr," Decuir growled piratically.

"Aye, you'd think that," Baumer said in her most nautical voice, "but it'd be the *sea*." Everyone groaned.

Since 2007, a rotating cast of Bridge Creek technicians have been tirelessly tracking the watershed's steelhead; sometime in 2017, they tagged fish number 100,000. That makes the project one of the most herculean research efforts in the Columbia River Basin, and one of its most revealing: Within its reams of data hides, perhaps, a secret vital to recovering imperiled salmon and steelhead runs. You've likely guessed that the secret involves beavers. But it also involves humans, and our ability to recruit rodents as allies. If Bridge Creek doesn't convince the fish world that beavers are integral to salmon recovery, nothing will.

———

The day before Jake Wirtz took me electrofishing, I toured Bridge Creek with Nick Weber, one of the architects of the stream's unconventional restoration strategy. Weber had the trucker's hat, patchy beard, and chill vibe of a fly fisherman, which he was, or a whitewater kayaker, which he also was. A near-death experience the previous day — his boat had gotten stuck in a hole on the White Salmon River, sending him through a frothy spin cycle — had only slightly harshed his mellow. "I thought that was it for me," he laughed.

Weber had worked on the Bridge Creek project since its inception in 2007, shortly after receiving his master's from Utah State University. Back then the project's research assistants pitched their tents for the summer atop a hellish patch of windswept earth that Weber not-so-affectionately called "Dirt Camp." These days Wirtz and his team live in a dorm-like bunkhouse endowed with a DVD player, giant refrigerators overflowing with eggs and sausage, and a front porch ideal for drinking beer and watching thunderstorms. It's not the Plaza, but it's not Dirt Camp, either.

Just as Bridge Creek's humans have experienced a habitat upgrade over the last decade, so have its fish. Weber led me into a wilderness of head-high willows, cattails, sedges, and beaver-sharpened stakes that threatened to impale the clumsy. If we'd been smart, we would have brought machetes. I quickly found myself locked in a tug-of-war with mousse-like muck, my sandals hanging in the balance. Once extracted, I caught up with Weber in a vast upstream beaver pool, where he waited in turbid, waist-deep water.

He gazed around like a college alumnus at a ten-year reunion, reminiscing about the days when Bridge Creek was more like Bridge Trench. "We used to walk up these banks and peer down into these straight steep channels that went almost dry in the summer, and be like, should we even be electrofishing this?" he recalled. "There was nothing there."

But that was then — before Weber and his colleagues installed more than a hundred so-called Beaver Dam Analogues, rows of posts and willows that transformed the stream's habitat, bolstered the survival of baby steelhead, and lured back beavers in numbers not seen since before the days of the fur desert. "This whole area went from moonscape to lush wetland," Weber marveled. "You're not squinting your eyes to see if maybe you made the habitat a little better. It's, like, 'Holy shit — look at that!' This may not look like a nice clear trout stream, but it produces steelhead like crazy."

To grasp exactly why building artificial beaver dams had such a salutary effect, it may help to endure a quick lesson in geomorphology. If you've driven around the American West, you've probably come across streams resembling the old, degraded Bridge Creek — straight gullies, cut deep into crumbling banks, nearly destitute of willow. These sorry creeks are victims of *incision*: the fatal disconnection of stream channel from floodplain. Trapped within their own channels, incised streams tend to erode to bedrock, like a spoon slicing through a scoop of ice cream until it scrapes the bottom of the bowl. Eventually their destabilized banks topple, and they become wider, shallower, and simpler. Over the years, they come to resemble nothing so much as the Los Angeles River, the pathetic trench immortalized by a thousand Hollywood car chases.

What could cause such rapid deterioration? Often the culprit is trapping. In a healthy, beaver-rich stream, dams and ponds act as step-like gradient controls that slow down flows and alleviate erosion, just as farmers might terrace a steep slope to prevent their topsoil from washing away. When beavers vanish and dams collapse, unchecked streams whisk away sediment and scour into gullies. In 1924, for instance, trappers eliminated beavers from Oregon's Crane Creek, where wet meadows once irrigated stirrup-high grasses. Just eleven years after losing its rodents, Crane Creek gouged out a deep canyon, and the grasses withered.[36]

Bridge Creek unraveled more insidiously. In addition to beaver trapping, the stream was also subjected to virtually unchecked cattle grazing.

Wallowing livestock mowed down streamside plants and trampled banks. Without "roughness elements" like willows and beaver dams checking Bridge Creek's velocity, the racing stream carved to bedrock. No longer could the stream spill onto its floodplain, energy dissipating as it spread like a puddle on a tabletop. With nowhere to go but downhill, Bridge Creek turned into a firehose.

To grasp the magnitude of Bridge Creek's collapse, put yourself in the scales of a young steelhead fry, no longer than a cigarette. You're aswirl in the current, hungry and skittish, hatched into a world of hungry beaks and maws. Even holding your position in the torrent is exhausting. You crave sluggish, debris-filled backwaters where you can gather strength and take cover from predators. But if your river has been turned into a featureless sluice, you're doomed to be blown downstream. You're screwed.

And that's if your stream still has water at all. When unbound rivers spill onto floodplains, the lingering moisture percolates through the soil and raises the water table, the subsurface horizon where soil becomes saturated. The higher the table, the more accessible the groundwater; in some well-soaked places, it actually seeps from the earth and mingles with surface water, a mixing process called *hyporheic exchange*. Scientists estimate that 40 to 50 percent of the water in many small and midsized streams comes from the ground.

Just as streams can receive water from the ground, they can also lose it as their flows dissipate into soil and rock. Streams that gain more water from the ground than they contribute to it are called *gaining streams*, while streams that lose more are called — you guessed it — *losing streams*. When creeks are too incised to soak their floodplain sponges, they can rapidly become losers as snowmelt and rainfall dwindle in summer. "You drop water tables so far that the stored water no longer exists," Tim Beechie, a fish biologist at the National Oceanic and Atmospheric Administration, told me. Beechie has found that terminal floodplain disconnection is depressingly common: When he studied the problem in western Washington, more than half the streams he examined suffered from incision.[37] "A lot of times you get a total loss of summer stream flows," he said. Incised streams aren't just subpar fish habitat, in other words: As the weather heats up, they're often no habitat at all.

Fortunately, there are ways to engineer a riverine jailbreak. Think of a stream channel as an uneasy truce negotiated by two competing processes:

erosion and its opposite, *aggradation*, the buildup of sediment. Streams whisk away silt, sand, and clay during their journeys downriver, but they also deposit this material, building up bottoms in some places even as they cut to bedrock in others. When streams slow down — on the inside of bends, at pools, on flatter terrain — they dump their loads, like a weary traveler dropping her baggage at the end of a long trip. You can observe this effect yourself: Next time you're alongside a muddy river, fill up a bucket and let it stand overnight. By morning, your pail of murky water will have divided into two distinct layers: clear water on top, and a few inches of sediment settled on the floor.

The best thing we can do for incised streams, then, is to tilt the scales toward aggradation by slowing down flows, helping creeks deposit their sediment loads, and allowing streambeds to rebuild themselves. Beaver dams — nature's speedbumps — can speed that process up. Tim Beechie found that beaverless streams in the Walla Walla and Tucannon River Basins would take up to 270 years to rebuild; throw in a handful of aquatic rodents, though, and you could shorten the recovery by a third.[38] The process applies beyond the Northwest: On the Piedmont plateau of North Carolina and Virginia, researchers estimated that beavers are capable of depositing twenty-two million cubic meters of sediment — the equivalent of twenty-one Empire State Buildings.[39]

That's the good news. The bad news is that the rebuilding process, left to its own devices, can take a really, really long time. After all, it's hard for a beaver to build a dam in a fast-flowing trench: A study that monitored Bridge Creek's beavers beginning in 1988 found that three-quarters of their dams didn't last more than two years, blown to smithereens by high flows. The harder it is for beavers to establish, the slower the rehabilitation — with potentially dire consequences for fish whose future depends on healthy habitat.[40]

The powerful hypothesis that underpins the grand Bridge Creek experiment, then, is this: When *Castor canadensis* can't repair a degraded stream on its own, *Homo sapiens* can give the rodents a leg up by imitating their works.

The Bridge Creek scheme was the brainchild of Michael Pollock, who has admired beavers since studying their habitat-creating prowess in Alaska in the early 1990s. When, a decade later, Pollock stumbled upon Bridge Creek, he recognized another ecosystem that stood to profit from beavers. In a 2007 study he found that some of the stream's dams captured enough sediment to raise the channel nearly half a meter a year.[41] "Beavers were restoring this incised stream, but their dams were blowing out constantly," Pollock told me. If a few collapsing dams were good, well, Pollock figured more stable ones would be better. He proposed an experiment that, to many fish biologists, seemed the height of insanity: he would supplement Bridge Creek's natural beaver structures by building dams of his own. "Nobody really understood it — 'You're going to drown riffles in silt, open up all this warm water, and on top of it you're putting in *dams*?'" Pollock laughed as he recalled his colleagues' incredulity. "'Do you know how long we've been trying to take out dams?'"

Pollock's idea didn't just raise biologists' hackles — it also seemed like a logistical nightmare. How could primates with puny teeth mimic the natural world's most talented builders? Pollock, along with an ecologist named Nick Bouwes, solicited artificial beaver dam designs from consulting firms. "The bids were coming back at fifty thousand dollars for one structure," Bouwes recalled. "I was appalled. I'd just gotten done building a log home for that much."

Bouwes — aka Big Nick, to Weber's Little Nick — combed the internet and found a thriftier alternative: a hydraulic post pounder, a handheld machine that resembles a cross between a jackhammer and a bazooka. Using their new toy, the Bridge Creek team installed 121 Beaver Dam Analogues between 2009 and 2012, pounding peeled logs into the creekbed and threading willow branches through the posts as though weaving a wicker basket. ("My back still hurts," Pollock told me.) The crew played with different sizes of BDAs, as well as different functions: Some were meant to capture sediment, others to scour out pools, still others to widen the channel by shunting flows around their ends. Erosion, in a sense, was their ally: If you're going to rebuild an incised channel, the sediment has to come from *somewhere*. The overarching goal was to convert a drastically simplified stream into a complex one. Or, as Weber put it, "We were just looking to use beaver dams to turn Bridge Creek into an absolute shitshow."

Humans may have opened the curtain on the shitshow, but it was beavers who performed the final act. "Wherever we put structures, beavers came and set up shop," Weber told me. Although many dam analogues breached (not necessarily a bad thing: Broken BDAs provided primo fish habitat), others provided a stable base of operations for expanding rodent populations. By 2013 beavers had built 115 dams of their own, and fortified nearly sixty BDAs. All told, Bridge Creek's beaver activity increased an astounding eightfold. What's more, many of the new dams sprang up outside the BDA construction zone, suggesting that beavers may have been drawn to the dam analogues, then dispersed throughout the stream. "One of my proudest moments was walking along Bridge Creek and saying, 'you know, there really ought to be dams in those three spots,'" Pollock said. "And the next year, boom, beavers have built dams exactly where we thought they should be." The implications were profound: Restorationists could steer beavers to sites in desperate need of help without going through the rigmarole of relocation.[42]

Thanks to the voluminous data collected by Jake Wirtz and his crew, we know, too, what a boon the Beaver Dam Analogue project has proved for fish. In 2016 the team published a blockbuster paper in *Scientific Reports* comparing the fate of juvenile steelhead in Bridge Creek to fish in another stream, called Murderers Creek, where the crew didn't build any BDAs. The results were unequivocal: Bridge Creek's mess of channels, pools, and slackwaters produced nearly three times more fish than the impoverished control stream, and steelhead survival rates were 52 percent higher. "Habitat changes that we thought would take a decade happened in one to three years," Bouwes told me. The study also detected adult steelhead navigating more than two hundred dams on their way to spawning grounds — putting the lie to the notion that beavers stymie fish passage.[43]

That wasn't the only myth they dispelled. Fish biologists often fret that beavers make creeks hotter; if that were true, it could spell dire trouble for cold-loving steelhead, who get stressed at seventy-seven degrees Fahrenheit and start dying at eighty-four degrees. The group found that far from boiling baby fish, however, beaver dams and dam analogues actually blunt daily heat spikes in summer, preventing temperatures from skyrocketing toward the lethal threshold — a blessing for steelhead.[44] Dams may pull off that feat by encouraging the all-important hyporheic exchange, forcing water underground to cool off before emerging through downstream gravel.

If any research effort is going to dispel the fish world's lingering anti-beaver prejudice, then, it should be Bridge Creek. "When we started this project, we had a huge amount of pushback," Weber told me as we clawed our way out of the jungle-like channel. "Then we got these publications out, and now every agency and nonprofit out here wants to build BDAs everywhere." He grinned wryly. "It just took a decade."

———

Thanks to the Bridge Creek project and its attention-grabbing research, mimicking and employing beavers has at last become one of the hottest trends in the salmon world. Two weeks after my visit to Oregon, I traveled south to witness a jaw-dropping version of the approach, applied to a basin even more screwed up than the Columbia.

Northern California's Scott River, a sixty-mile tributary of the Klamath that gushes out of the Shasta-Trinity National Forest, less resembles a natural watercourse than an environmental crime scene. On a sweltering June morning, Charnna Gilmore led me among towering gravel embankments, stories tall and hundreds of yards long, that snaked across the valley like oversized gopher tunnels. The meandering slagheaps lent the floodplain the appearance of a post-industrial sacrifice zone.

That's essentially what the Scott Valley is. Beginning in 1908, a procession of monstrous, barge-like gold-mining dredges gouged the riverbed and belched out their tailings along the banks. "During the first twelve months in operation [the dredge] turned over 7.5 acres of ground and handled 354.961 cubic yards of gravel, while digging to an average depth of 30 feet," California's state mineralogists wrote in 1910, sounding more awed than alarmed.[45] Gilmore, executive director of the Scott River Watershed Council, was less impressed. "This," she said, as our sandals crunched over rubble, "is what we call 'completely hosed.'"

Once, the gravelly dumping ground beneath our feet would have been a verdant marsh, its pools and sloughs packed with fish and shaped by rodents. Beavers abounded in the Scott, slowing flows, filling wetlands, and opening side channels; the animals were so prolific here that early colonists called the region Beaver Valley. That abundant habitat would have been especially important for coho fry, who spend longer in their natal rivers than the offspring of other salmon species. But the beavers didn't last. In 1850 the mountain man Stephen Meek, a shrewd Virginia native who resembled Tolstoy in buckskins, hauled out eighteen hundred pelts in a single year. Eighty years later, in the winter of 1929, a trapper named Frank C. Jordan extracted the Scott's final beavers.

If Charnna Gilmore has her way, the Scott River could become Beaver Valley once more. In 2014 the watershed council, with design help from Michael Pollock, installed a total of eight artificial beaver dams on four Scott tributaries. Gilmore took me to Sugar Creek to see one of their creations, a much more substantial structure than the brushpiles I'd seen at Bridge Creek. A row of locally sourced Douglas fir logs jutted from the creekbed, interwoven with a lattice of willow branches and buttressed with rocks; the whole wall stretched longer than a bus. Behind the dam, a sprawling pond, broad as several end-to-end tennis courts, backed into the alder. Silver fry flitted around half-sunk clumps of willow. Gilmore slogged through the pond, the water lapping at her thigh-length shorts, rejoicing at every feature: the chocolatey wedges of sediment, the beaver-dug canals, the chewed sticks with which rodents had fortified the dam.

"I think he — or she — has found a partner," she said. "There's a lot more scenting. I'm hoping they're having babies."

Gilmore, a tireless, loquacious woman who wears her dirty-blond hair tied beneath a baseball cap, was hardly destined to become a Beaver Believer. A native of central Washington, she moved to the Scott Valley after high school to maintain trails for the Forest Service and never left, selling real estate for eighteen years. "There's a hardness about this place that builds character," she told me. In 2007 she joined the board of the watershed council and fell in love with ecological restoration; eight years later, still working as a real estate agent, she enrolled as an undergraduate in the College of the Siskiyous. (She also joined the cross-country team, inspiring the *Siskiyou Daily* to write a glowing feature about the squad's forty-seven-year-old freshman.) In 2016 she finally abandoned her real estate career to run the council. She might be the only nonprofit director in the country still pursuing a bachelor's degree in environmental science. "My kids think I'm nuts," she said merrily.

Although Gilmore and her council colleagues long yearned to restore beavers to the Scott, they feared antagonizing its conservative ranchers and loggers. "I would've never worn a T-shirt with a beaver on it in Scott Valley five years ago — no way," said Gilmore, who was now wearing exactly that. "We were like closet beaver people. We didn't want to start seeing bumper stickers saying, you know, WIPE YOUR ASS WITH A BEAVER."

The antagonism began to soften in 2012, when California was stricken by drought. In some corners the five-year dry spell was the worst to hit the state in 450 years. The crisis didn't go to waste in the Scott, as beaver ponds elevated water tables, reduced irrigators' expenses, and converted former foes. Thanks to beavers, a rancher named Gareth Plank told *OnEarth* magazine in 2015, "We've had a 10 percent to 15 percent reduction in pumping costs."[46] Far from opposing the watershed council's artificial dams, property owners began to clamor for them. (Despite the growing goodwill, beavers remained controversial enough that the council gave their dam analogues the most benign name imaginable: Post-Assisted Wood Structures.) Within a year, some of the PAWS had raised water tables between one and three feet.

"The landowners who want us to put these up aren't really interested in fish," Gilmore told me. "They just want to talk about water storage and irrigation."

The most obstinate resistance to the Scott Valley structures came not from local ranchers, but from the state of California. The Department of Fish and Wildlife limited the group's placement of dam analogues and, when drought grew especially severe, tried to force their removal for the sake of fish passage. While wildlife agencies in neighboring states like Oregon and Washington have embraced beavers, California has been notably hesitant, belying its otherwise sterling reputation as an environmentally progressive state. "Elsewhere, a lot of these techniques are well documented, well accepted," Kate Lundquist, a beaver advocate at the Occidental Arts and Ecology Center, told me. "But working with beaver hasn't been recognized as a legitimate tool in California for a long time."

Gilmore told me her group's relationship with the department was "evolving" — two weeks before my visit, the Scott River Watershed Council invited state biologists to a dam analogue workshop intended to dispel anxiety. But she recognizes she's battling prejudices with century-deep roots. Few places stand to benefit as much from beavers as California, yet no state quite shares its strange history of beaver denialism. "There's this dichotomy between what we believe the beaver to be," Gilmore told me, "and how the state has historically seen this animal." The future of the Golden State's beavers may depend on how California reckons with its past.

—CHAPTER SIX—

California Streaming

To my knowledge, the world's largest collection of beaver-themed tchotchkes, knickknacks, and memorabilia is housed on an oak-shaded street in Martinez, California, in a home whose front porch is guarded by a mural the size of a picture window — a reddish beaver, stick grasped in forepaws, tail raised in salutation. The dim interior has the feel of a shrine. Beaver magnets cling to the refrigerator; plush beavers perch atop the bureaus; a gallery's worth of beaver paintings, prints, and posters stare down from the walls. Gnawed stumps rest next to the fireplace. Embroidered beaver napkins hang in the kitchen. In the backyard, a clay beaver crouches in the birdbath. If I'd come during Christmas, I would have seen a cardboard beaver cutout, as large as a black bear, strung with lights on the front lawn.

The curator of this collection is a candid, vivacious woman named Heidi Perryman, a child psychologist who, through willpower and single-mindedness, has become one of the planet's foremost authorities on *Castor canadensis*. After months of exchanging emails, Perryman and I resolved to rendezvous in Martinez on the Fourth of July. She prefaced our meeting by sending me a link to her own avowal of patriotism, a YouTube video she'd created called the Beaver Pledge: *One river, underground, irreplaceable, with habitat and wetlands for all.*

On Independence Day, as the rest of Martinez paraded through downtown in their best red-white-and-blues, I met Perryman and her husband, Jon, for brunch at their home-*cum*-museum. Outside, tiny beaver silhouettes were etched in the sidewalk, the fingerprint of a rodent-shaped cookie cutter pressed into wet cement. No sooner had I arrived than Perryman loaded me up with souvenirs, including a family of clay beavers mounted on a bamboo stand and a lime-green tie emblazoned with dozens of fingernail-sized you-know-whats. "Did you notice your coffee?" she asked slyly, after I'd taken a few sips. I glanced down to lock eyes with a ceramic beaver squatting like a toad at the bottom of my mug.

Ask a fellow Beaver Believer to characterize Heidi Perryman, and the primary descriptor you'll hear is *force of nature*. Perryman's main endeavor is Worth a Dam, a nonprofit that serves as a comprehensive clearinghouse for beaver science and coexistence techniques; a beaver news outlet, updated daily; a beaver-themed LinkedIn that connects global Believers with resources and one another; and a sort of gossip blog for the castor cognoscenti. (From a post that appeared the morning after my visit: "Yesterday was a fun and oddly familiar day . . . with very busy author Ben Goldfarb who laughed often at the story, took notes on a little pad, and recorded the interview with his phone.")

If you are even tangentially connected to the beaver world, you have crossed paths with Heidi Perryman. She is a prolific poster in beaver-themed Facebook groups and an indefatigable disseminator of information; many times I have woken to find some beaver-related tidbit in my inbox she'd gleaned from Alaska, Alabama, or Scotland. Nothing escapes her attention — nor, at times, her scorn. After a popular nature podcast ran a beaver episode that she considered biased toward lethal control, she responded on her blog with a take-no-prisoners diatribe: "I don't know about you but I've reached my CULTURAL CARRYING CAPACITY for stupid-ass reporters like this who repeat beaver bullshit even though they have the real answers RIGHT at their fingertips."[1] Her polemics are also, it must be said, highly effective: After Perryman sent the offending podcast an annotated transcript with its errors highlighted, the host promptly ran a contrite and even-handed follow-up.

"As a psychologist, I have to be so thoughtful about what I say," she told me as we ate quesadillas on the back porch. "It comes as a relief to rant on the website once in a while."

Perryman's fervor is all the more remarkable considering that, when beavers arrived in downtown Martinez in 2007, she didn't know the first thing about the newcomers. Truth was, she'd seen only a single beaver in her life, a lone adult that tail-slapped as she and Jon canoed in Mendocino County. Her naïveté extended throughout the realm of aquatic mammals: When she took a picture of a brown critter perched atop a new lodge in Martinez's Alhambra Creek, she didn't realize until later that she'd mistakenly photographed an otter.

Still, you could forgive Perryman her ignorance. At the time, hardly *anyone* in the Bay Area had seen a beaver. California may otherwise be a mecca of environmentalism, a honeyed land of plastic bag bans and cap-and-trade programs, but when it comes to beavers the Golden State is decidedly retrograde. In large part that's because it has bent nature to its will like no other place. California, according to the Water Education Foundation, is "the most hydrologically altered landmass on the planet,"[2] its waters choked by dams, impounded by reservoirs, shunted to and fro by a sprawling circulatory system of aqueducts and irrigation channels that keep taps flowing in Los Angeles and crops growing in the Central Valley. The state's fundamental dilemma is that its urban centers and fertile soils don't quite overlap with its most bountiful hydrological resources; it has solved that problem — or attempted to solve it — with pumps, canals, and concrete.

"It is easy to forget that the only natural force over which we have any control out here is water, and that only recently," wrote the essayist Joan Didion. "In my memory California summers were characterized by the coughing in the pipes that meant the well was dry, and California winters by all night watches on rivers about to crest, by sandbagging, by dynamite on the levees and flooding on the first floor. Even now the place is not all that hospitable to extensive settlement."[3]

California's victory over the natural world has been hard-won, and its reluctance to give quarter to interfering rodents is perhaps understandable. But the state's story is more complicated than a simple failure to coexist. It's about the geography of western exploration, the historical limits of memory, and a very consequential error by an astute biologist. It's about how imperfect knowledge can become enshrined in decision making, and how gaps in understanding can lead wildlife management astray. On top of it all, it's about how the most hydrologically altered landmass on earth began, falteringly, to welcome back North America's most meddlesome mammals.

Heidi Perryman did not set out to change the fortunes of California's beavers. When, back in 2007, the first pair began building in Alhambra Creek, she was simply delighted by the novelty. "They were adorable," she told me, before revising her opinion. "Well, they were unusual. They were more unusual than adorable. Actually, they're *not* really that adorable — but they were very cool." Perryman was most enamored of the life that rode in on the beavers' coattails: herons, otters, mink, muskrats. She and Jon strolled daily down to the bridge that spans Alhambra Creek to film the frolicsome creatures. More than a decade later, she has external hard drives loaded with two terabytes of beaver footage — the equivalent of around a dozen MacBooks' worth.

The city of Martinez, however, was less enchanted. Alhambra Creek flows through downtown on its way to San Francisco Bay; during heavy winter rains, the stream is prone to rampaging through the streets. Although Martinez alleviated the problem with a ten-million-dollar flood control project in 2001, the specter of deluge still loomed large. The town wasn't sure whether beavers represented a true threat, but creek-abutting business owners preemptively complained. The Martinez city council reassured its constituents that the beavers would be killed.

The announcement alarmed Perryman, who'd fallen head over heels. The beavers had recently birthed four kits, who actually *were* adorable, and who uttered the most beguiling squeaks and gurgles. "I remember thinking, do the people that want them killed even know about the sound that a baby beaver makes?" Perryman said, the silver beaver pendant on her necklace glinting in the sun. "And if I don't do something, will I ever hear that sound again?"

At this point in our conversation, Perryman decided her story required a visual aid. "Jon!" she hollered toward the interior of the house. "Bring the scrapbook! Oh, and could we have more coffee? Some waitress you are." A moment later, Jon, a genial fellow who wore a WORTH A DAM tank top and his hair in a silver ponytail, emerged with a swollen scrapbook, its pages bursting with the paper trail of Perryman's campaign. I leafed through the documentary evidence of her struggle: pre-stamped, pro-beaver postcards she'd handed out to pedestrians on the bridge; articles she'd written for the *Martinez News-Gazette*; lyrics to a Blue Oyster Cult parody song ("City don't kill the beavers"). The *San Francisco Chronicle* and *Los Angeles Times* covered the quirky controversy. The city announced that the California Department of Fish and Wildlife would live-trap the animals and relocate them to tribal land, assuming this would mollify the beaver freaks. The freaks were not mollified. Schoolchildren stood on the bridge and chanted "Leave her, leave her, save Ms. Beaver!" Even the city's Wikipedia page became a hotbed of dispute, a battleground between editors praising Martinez's beavers and anonymous trolls castigating them.

Martinez already possessed something of a split personality: Its two most famous landmarks are the former residence of John Muir and a foreboding Shell Oil refinery. The beaver brouhaha only deepened divisions. "It was a Hatfield and McCoy scenario — either you were totally for beavers or totally against them," Mark Ross, the city council's lone pro-beaver member, told me. Some of the community's wealthiest and most powerful pillars were vehemently opposed. During one confrontation, a well-heeled businessman cursed in Ross's face, their noses so close they practically touched. "I was thinking, *This seventy-year-old guy is about to hit me!*" Ross recalled. "*Do I hit back against a senior citizen or not?*"

At last, the worn-down city agreed to hold a public meeting. On November 7, 2007, two hundred people packed into a high school auditorium. Eleven police officers had been summoned to mind the restless crowd. The first person who stepped to the mike demanded the city remove the beavers. The next forty-nine demanded they stay. Tim Platt called the beavers the best thing to happen to downtown Martinez in years. Katherine Myskowski and Linda Aguirre said they were tourist attractions. Sheri-ann Hasenfus claimed they'd brought the city together. Charles Martin suggested the high school change its mascot from bulldog to beaver. The mayor read a comment card from a nine-year-old girl named Natalie who feared for the beavers' future.[4]

The city council did what city councils do: It created a subcommittee. Reluctantly, they gave Heidi Perryman a seat.

The battle lurched on into 2008. Perryman, finding the internet lacking, scoured the country for beaver-smart professionals to advise the city, stumbling finally upon Skip Lisle and Beaver Deceivers International. At Perryman's recommendation, the city spent $10,500 flying Lisle from Vermont to Martinez to install a Castor Master. The *News-Gazette* commemorated the event by running a front-page photograph of Lisle mucking around in the pond, bare arms rippling in the sunshine, below the headline "Burly Beaver Biologist Breaks a Sweat." A yellowed copy of the article — signed *BBB* by the burly beaver biologist himself — is pressed into Perryman's scrapbook. "I've never had media coverage like that," Lisle marveled to me. "Every news outlet in San Francisco seemed to be there."

Unlike so many beaver tales, Perryman's concludes happily: The Castor Master worked. Alhambra Creek didn't flood. The city never removed the beavers, but they never quite countenanced them, either. Skirmishes occasionally flared: When, in 2011, an artist named Mario Alfaro painted a beaver into his mural celebrating Martinez's history, the city made him erase it, like John D. Rockefeller demanding Diego Rivera excise Vladimir Lenin.[5] (Alfaro got the last laugh — if you look closely at the mural today, you can spot a little leathery tail descending from the final *O* in his signature.)

The years passed. The beavers stuck around, integrated into civic life, another ingredient in the urban melting pot. Even detractors moved on, though Mark Ross, the pro-beaver councilman, told me some business owners still don't talk to him. Beyond its borders, Martinez gained a reputation for castor activism. "To this day, if you go around to obscure corners of the Bay Area and tell people you're from Martinez, they'll go, 'Oh, how are the beavers?'" Ross told me. "'I'm so glad you didn't kill them.'"

The only person who continued to live and breathe aquatic rodents was Heidi Perryman: Once a Beaver Believer, always a Beaver Believer. In 2008 Perryman held the first-ever Martinez Beaver Festival, a quaint affair with eight booths and a few cardboard tails on which kids could glue stickers. "We thought it would be harder for the city to kill them after we'd thrown a party for them," she told me. Within a few years it had become one of the most beloved events on the city's social calendar. Worth a Dam, whose website she'd built with programming help from a local homeless man, took off, too. Every week, it seemed, another email drifted in from another beaver-lover seeking advice about how to save their local colony from heavy-handed managers.

"When I first started doing this, I thought maybe Martinez was uniquely stupid, and that I'd read all these stories about other cities being smarter than us," Perryman told me. "But it turned out that Martinez was just run-of-the-mill stupid."

None of the foolishness she encountered, however, held a candle to the faulty assumptions that underpinned her own state's approach to beaver management.

———

In 1987, soon after buying a house in Silicon Valley, a doctor became obsessed with a stream. The doctor was Rick Lanman, a cancer geneticist; the stream was Adobe Creek, the seasonal watercourse that ran behind his new Bay Area home. Although the stream went dry in summer and fall, Lanman learned from the previous owner that Adobe Creek — the inspiration, in true Silicon Valley fashion, for the name of Adobe Systems — had once flowed year-round. The former owner recalled fly fishing for steelhead in his backyard. That sounded pretty good to Lanman. But where, he wondered, had the water gone?

Lanman came to suspect that Adobe Creek had, somewhere along the line, lost its beavers. The theory made intuitive sense, but it was contradicted by one inconvenient fact: According to the California Department of Fish and Wildlife, beavers weren't native to the Bay Area. In fact, the agency averred, the rodents had originally been absent from most of coastal California. Nor had beavers historically occupied the Sierra Nevada range, the state's granitic spine. Any beavers that dwelled in those places today were the unnatural scions of human releases — not the descendants of natives.

Most people would have accepted the state's explanation and forgotten the matter. Rick Lanman is not most people. Lanman is an avid naturalist, the intellectual heir of a centuries-long line of physician-collectors, from William Tudor to Hans Sloane. Where those men might have pinned insects, Lanman takes a modern approach to his craft: He is happiest combing digital archives and writing Wikipedia entries for species and geographic features. Although he's widely published in the medical literature, sometimes it seems he pours as much energy into his avocation as his career. The first time we spoke, I called him from New Haven, Connecticut, near a brackish tendril of the Long Island Sound called Mill River. Lanman, three of whose sons had attended Yale, mentioned casually that he'd written the Mill River's Wikipedia entry, as well as the nearby Quinnipiac River's, not to mention those for a handful of other

New Haven rivers and creeks. If you visit the Wikipedia page for *Castor canadensis*, most of the words you read will be his.

To Lanman, the claim that beavers didn't belong in his stream wasn't the final word — it was a thrown gauntlet. And the deeper the autodidactic doctor researched beavers' natural history, the more preposterous the state's position seemed. "It just made no sense," he told me. "Beavers thrive from the tundra line in Canada to the Sonoran Desert in northern Mexico. What could keep possibly them out of most of California? How could our climate be so fiendish?"

Lanman soon came to realize he wasn't the only Californian plumbing beaver history. In 2009 he attended a Silicon Valley water conference whose keynote speaker was Brock Dolman, an ecologist at a Sonoma County nonprofit called the Occidental Arts and Ecology Center. Dolman had been touting beavers in his lectures at salmon biology meetings since 2007 to tepid response. "I mostly got laughed out of the room, heard a couple beaver innuendo jokes, the usual stuff," Dolman told me. His talk at the Silicon Valley conference, however, elicited a different reaction, at least from one attendee. "Afterward this excited guy comes up and wants to talk beaver."

The excited guy, of course, was Rick Lanman. The ecologist and the doctor — soon joined, inevitably, by Heidi Perryman — formed a de facto beaver detective agency, scouring the internet for rodent-related information and exchanging their discoveries in flurries of emails. As it turned out, there was plenty of material for three beaver obsessives to unearth — especially when it came to the animals' native range in the state of California.

As the gumshoes perused Indian place-names, trappers' journals, newspaper accounts, and other sources, the evidence for beavers' historical ubiquity mounted. Along California's coastline, every native tribe seemed to have a word for beaver: The animals were *kat-si-keh'* to the Round Valley Pomo of the Eel River, *kah-ka'* to the Coast Miwok of Marin, *'eveeenxal* to the Luiseño of San Diego. Beaver teeth appeared in the Emeryville Shellmound and a midden in Humboldt County. An 1827 diarist mentioned native people at Mission Sonoma "letting fly their arrows at the beaver."[6] In the Sierras, where streams were supposedly too steep and flashy for beavers to ascend, trappers' journals reported "plenty of Beaver," and five-hundred-year-old beaver paintings adorned the Tule River reservation.[7] Geographic references to Beaver Meadows, Beaver Creeks, and Beaver Ranches abounded. "I'm *still* finding stuff," Lanman told me.

How, then, had California so badly misinterpreted its rodents' range? The beaver detectives, rather blasphemously, suspected the fault lay with one of the state's most revered biologists: Joseph Grinnell.

If you've ever taken a high school biology class, you are likely familiar with Grinnell's work, even if you don't know his name. Born in 1877, Grinnell was a disciplined field biologist who, among other contributions, developed the concept of the *niche*, the notion that each species fills its own distinct ecological role. Grinnell traveled throughout California, meticulously mapping the distribution of the state's birds and mammals in voluminous notebooks. His reports were so detailed and reliable that modern scientists still repeat his surveys to track how climate change is altering wildlife habitat.[8] He served as the first director of Berkeley's venerable Museum of Vertebrate Zoology, and preached conservation biology before the term even existed, advocating for the protection of predators in an era when many scientists viewed destroying wolves, coyotes, and mountain lions as a holy crusade.

When it came to beavers, however, the great Grinnell missed his mark. In 1937, two years before his death, the zoologist published an epic volume entitled *Fur-Bearing Mammals of California*. Grinnell's section on beavers was, as per his usual, exhaustive and perceptive. On many ranches, Grinnell wrote, "the building and maintenance of dams by beavers and the resultant raising of the water table has been of decided benefit." Yet Grinnell woefully misdiagnosed the species' native range. Despite the Sierra Nevada's prime alpine streams and hardwood groves, Grinnell claimed it had historically been bereft of beavers, which "never existed at an altitude higher than about 1000 feet." Even more confounding, he claimed beavers originally dwelled only in the Central Valley; the Pit and Klamath Rivers, in far northern California; and in the Colorado watershed, at its southern rim. Most of the state — including the Bay Area, the Sierra Nevada, the entire coast, and virtually all of Southern California — had, according to its foremost biologist, been beaverless.[9]

Grinnell's conjectures took stronger hold in 1942, when a successor named Donald Tappe published a monograph about California beavers. Tappe acknowledged that the Sacramento and San Joaquin Rivers, the twin waterways that drain the Central Valley, had once been rife with rodents. But he co-signed Grinnell's claim about the Sierra Nevada's barrenness, and speculated that the innumerable salmon streams that edge California's coastline had also lacked beavers — the southern streams because they were too dry, the northern streams because they were "rocky and steep with but little beaver food growing along them."[10]

Even to their creators, the maps seemed perplexing. Grinnell called it "curious" that the Sierras lacked beavers, even though "mountain streams and bordering growths of deciduous trees abound, like those which constitute ideal abodes of beaver in Colorado and elsewhere."[11] But the dubious assumptions became gospel, unchallenged by succeeding generations of scientists. When the California Department of Fish and Game published a review of the state's mammals in 1986, for instance, it listed the beaver's altitudinal range as "sea level up to about 1000 ft" — never mind that beavers live in places that are drier, rockier, steeper, and higher all over the continent.[12]

What Grinnell and Tappe had failed to account for was history.

———

In many ways California's beavers followed a fairly typical trajectory. Jedediah Smith led trapping parties to the region in the late 1820s, becoming the first white Americans to reach California through the Southwest, the first to travel north along the coast to the Columbia River, and the first to traverse the Sierra Nevada and the fearsome "Sand Plain" beyond.[13] Peter Skene Ogden's ruthless band ransacked streams around Mount Shasta, Klamath Lake, and the Humboldt River. Those pioneering journeys opened the state to still more trappers and traders, who found a furry bounty waiting for them in the marshy Sacramento Valley.

But in one key respect, the California story diverged. Along the Pacific Coast, unlike the rest of the continent, beavers were not the most valuable pelt on the market. Sea otters, lovable maritime cousins to weasels, gamboled from the Aleutian Islands to the Baja Peninsula, prying sea urchins, crabs, and bivalves open upon their plush bellies. To survive the frigid Pacific and their own deficit of body fat, otters evolved the animal kingdom's thickest pelt — up to a million hairs per inch, several times denser than even the beaver's underwool. When Russian explorers returned from

Alaska bearing exquisite otter pelts in the 1740s, the *promyshleniki*, a class of professional hunters whose brutality surpassed even the Americans', sprang into action. Boatloads of *promyshleniki* ravaged the Aleutian Islands, rounding up natives, forcing them to slaughter otters, and massacring dissenters. Hunters shipped "soft gold" to China and turned their attention south, toward California. Meanwhile Captain James Cook stumbled upon the lucrative trade during his West Coast voyages, and his logbook, published posthumously in 1784, stirred American and British merchants to action. By 1801 a steady crawl of merchant ships, most from Boston, were braving the tempestuous churn of Cape Horn en route to the West Coast, where they loaded their holds with furs before sailing to China.[14]

Although otters were the targets of the West Coast trade, plenty of beavers met their ends as well. While trading for sea otters and fur seals along the California coast and the Columbia River, for instance, the ship *Albatross* also scooped up 248 beaver pelts. The Russian ship *Kodiak* collected beavers as well as otters from Bodega Bay before heading back to Alaska in 1809.[15] Just as native trappers once funneled pelts to English and Dutch traders on the East Coast, California's coastal tribes piped furs to American and Russian mariners throughout the late eighteenth and early nineteenth centuries — years before Smith, Ogden, and other overland parties showed up. California's wildlife was also exploited by the Spanish, who'd occupied the region since 1697.

The upshot is that, by the time overland travelers reached California from North America's interior, the state's beavers had already been depleted. Beaver-minded explorers, Jedediah Smith among them, reported finding the animals scarce in many coastal and Bay Area streams, unwittingly reflecting a century of trapping and trading. Streams that spilled from the Sierra Nevada into the Central Valley were trapped top-to-bottom by Smith and his ilk, but American trappers kept few records of their pickings. What's more, those few beaver specimens that might have been collected by early zoologists didn't endure. In 1906 a 7.8-magnitude earthquake rocked San Francisco, burning the California Academy of Sciences to the ground and incinerating the state's only pre-twentieth-century zoology collection.

Conservation biologists refer often to the notion of *shifting baselines syndrome*, a form of long-term amnesia that causes each successive generation to accept its own degraded ecology as normal.[16] Salmon fishermen who rejoice at catching ten-pound chinook forget that their fathers once hauled out fifty-pound behemoths. Modern biologists who marvel at mayfly hatches never saw the insects emerge in clouds so thick their bodies piled up in three-foot-deep windrows. Every year our standards slip a little further; every year we lose more and remember less.

Joseph Grinnell and Donald Tappe may have fallen victim to shifting beaver baselines themselves. The notion that California was beaver-poor — a historical artifact created by the sea otter trade — became enshrined in books and conventional wisdom, with grim consequences. When, for instance, beavers felled cottonwoods and willows at a nature reserve outside Los Angeles in 1998, managers ordered them trapped, ostensibly for the sake of protecting nesting habitat for endangered Bell's vireos and southwestern willow flycatchers. It was, of course, a backward decision: Researchers later found that officials had ignored "best available science," including a hefty body of literature suggesting that beavers create prime songbird habitat.[17] The reserve's administrators, needless to say, were far from the first to make poor beaver decisions. Still, in their insistence that the engineers would harm indigenous songbirds, it's easy to detect anti-beaver nativism: The rodents simply didn't belong in Southern California, or in many other parts of the state.

That hostility might have persisted indefinitely, were it not for Charles Darwin.

Not *that* Charles Darwin: The one in this story is Charles Darwin James, a genial former US Forest Service archaeologist who goes by Chuck. (Chuck is actually Charles Darwin James III, although his family, so far as he knows, is devoid of evolutionary biologists.) Back in 1988 James's colleagues told him they'd noticed some old beaver dams in Red Clover Creek, a tributary of the Feather River that begins as a trickle through an alpine meadow fifty-three hundred feet up the Sierras — unremarkable, except for the fact that, according to Joseph Grinnell, they weren't supposed to exist that high. When James visited Red Clover, he found that massive flooding two years earlier had cut deep into its banks, exposing long-buried jumbles of sticks chiseled by characteristic toothmarks. A year later James returned to the meadow and, with a trowel, extracted sticks from three ancient dams. He wrapped the samples in tinfoil and shipped them off to a lab in Miami for carbon-dating, which pegged the specimens' origins at 1850, 1730, and

AD 580. In direct contradiction to Grinnell's historical range maps, beavers had occupied the High Sierra for at least a thousand years before the arrival of white people.[18]

Although James never published his findings, he told me he'd mentioned the dams to the California Department of Fish and Wildlife, to no avail. "People who've invested a chunk of their careers championing one theory or another are pretty resistant to change," he said. Two decades later, though, Heidi Perryman, beaver networker extraordinaire, caught wind of James's story and connected him to Rick Lanman, who realized his new acquaintance might have found a smoking gun. In 2012 James and Lanman coauthored twin papers — one about the Red Clover Creek dams, the other, with Brock Dolman and Heidi Perryman, an exhaustive review of Indian place-names, observer records, and additional evidence that beavers inhabited the Sierra. A year later the sleuths, now joined by a team of colleagues that included Lanman's own son, published a parallel study proving that beavers had inhabited the same coastal streams that Tappe once claimed were too dry or rocky to support them. Lest their work escape the state's attention, the team published its research in the California Department of Fish and Wildlife's own journal. No member of the group felt quite so vindicated as Charles Darwin James, the archaeologist whose astonishing Sierra discovery had been dismissed for thirty years. "It wasn't so much personal pride — I was just happy to be part of something that's bigger than myself," he told me. "And I was most happy for the beaver."

Together the group's three papers formed an airtight case: Beavers had dwelled nearly everywhere in California, with the possible exception of the arid Mojave. Lanman, invigorated, told me he's revisiting the state's historic range maps for even more contentious creatures, like wolves and jaguars. "In medicine we believe stuff that's wrong just because some eminent guy said it, and no one questions it for years," he said. "The same is true of historical ecology: You have to question current beliefs in order to guide accurate restoration."

In the end Fish and Wildlife never publicly announced beavers' revised status — though they didn't deny the evidence, either. "The state's in a very ambivalent situation with beavers now," Heidi Perryman told me when we spoke in Martinez. "They're better than they were: They're no longer saying that they're not native. But they're definitely not saying they're not a pest, either." Old management habits die hard, in other words — especially in the nation's most hydrologically modified state.

———

In the summer of 1846, a militia composed of American colonists captured the city of Sonoma, declared their independence from Mexico, and hoisted the flag of the new sovereign republic of California: a red star and a red grizzly bear stamped against a white field. The Bear Flag Revolt was short-lived — American forces quickly claimed the region and replaced the bruin with the Stars and Stripes — but its symbol survived: Today the grizzly adorns the California state flag, and you can hardly travel north of San Francisco without meeting a dude rocking a Bear Flag cap. More than 160 years after the Bear Flag fell, I drove up the coast from Martinez to Sonoma County, through golden chaparral and redwood glades, to visit the paradisiacal headquarters of a smaller mammal-themed uprising — the Bring Back the Beaver Campaign.

The Occidental Arts and Ecology Center, the informal seat of California's Castor Revolution, steadfastly resists categorization. The center, eighty dappled acres of rolling orchards and oak woodlands, is part educational farm, part research site, and part commune, the kind of blissful agrarian utopia that could've been dreamed up by Gary Snyder and Wendell Berry after a long meditation session. When I arrived, the bushes were laden with gooseberries, raspberries, and blackberries, and apples, plums, and pears dangled invitingly from the trees. Gray fox kits scampered through the gardens, lapping at drip-irrigation hoses.

Even within the eccentric sect of Beaver Believers, Brock Dolman, one of OAEC's founders and Rick Lanman's conspirator in historical research, stands out in his idiosyncrasy. The morning after my arrival, Dolman took me on a tour of the sprawling property, curls flowing from beneath his Beaver Flag baseball hat. Dolman is an erudite naturalist possessed, as one colleague put it to me, of an "infectious biophilia"; as we walked, we stopped constantly to caress hawk tailfeathers and examine dead shrews, shuck seedheads and sniff bay leaves, pick through owl pellets and pluck mulberries. ("Just give it a tickle from below, a little scootchy-scootchy," Dolman advised as I reached for the thimble-sized fruit, "and those babies should fall right off.") To converse with Dolman is to stroll spellbound through a blizzard of alliterative aphorisms, elaborate riffs, and lyrical catchphrases, all sprung from his own singular mind. He doesn't talk about sustainable water management — he demands "a reciprocal rehydration regeneration revolution." California's hydrology is so constrained by metal and concrete that he prefers "pipeshed" to "watershed." And why call for diminishing the velocity of runoff when you can recommend "slowing, spreading, sinking, storing, and sharing"?

Like many Beaver Believers, Brock Dolman found religion through fish. Dolman is a certified salmon fanatic, a man so committed to preserving California's dwindling runs that he once attended a county meeting in a home-sewn coho costume and angrily spawned orange pom-poms across the desk of a conservation-averse commissioner. "If you have salmon in your basin," he told me as we nibbled tart plums, "every bit of land-use that you do is either salmon restoration or salmon destruction." In Sonoma County the scales have long been tilted toward destruction, in large part by rampant development and the countless acres of impenetrable concrete that have replaced permeable soil. Lacking opportunity to sink into the earth, rainfall rushes downhill over hard surfaces and through pipes, hastening erosion, exacerbating flooding, and sweeping whatever pollutants lie in its path into streams. Dolman's solution, concocted in the mid-2000s, was a philosophy he dubbed "conservation hydrology" — cooperating with nature, rather than piping and paving it, to recharge groundwater, alleviate floods, and filter runoff. The Occidental Arts and Ecology Center — whose own land drains into Dutch Bill Creek, a salmon-bearing tributary of the Russian River — became a conservation hydrology proving ground, blossoming with bioswales and roof catchments and gray water recycling systems; during my visit Dolman was especially keen to show me the new composting toilets processing "humanure." In Dolmanese, the key to saving salmon is transitioning from the "drain-age" to the "retain-age." And nothing, he has come to realize, retains water quite like a beaver.

Dolman's beaver cheerleading initially met a chilly reception. California, he discovered, suffered from a "beaver blind spot," the unfortunate legacy of Grinnell's error. As it does most rodents, California considers beavers a "detrimental species" whose transport could damage agriculture or public health, making it impossible for any group other than the Department of Fish and Wildlife to relocate one. (Contrast that with Washington, where nonprofits, fisheries groups, and tribes are all engaged in beaver relocation.) Kate Lundquist, the co-leader of the Occidental Arts and Ecology Center's Bring Back the Beaver Campaign, told me some state offices give away permits to kill beaver to just about any landowner who asks. Recreational trapping is legal in forty-two of its fifty-eight counties, and there's no limit to the number of beavers a trapper can take.[19] As Charnna Gilmore discovered, applying for permits to install Beaver Dam Analogues can be nightmarish. While other states treat beavers as habitat-creating, water-storing boons, California has historically regarded them as little more than vehicles for furs and irrigation-fouling pests.

"That department is schizophrenic," one former county water commissioner told me of Fish and Wildlife. "They have all these anadromous fish people giving speeches and writing articles and saying how wonderful beavers are. And then they have all these depredation types saying, Ah, to hell with the beavers, they cause all these problems."

Occidental's grassroots Bring Back the Beaver Campaign has brought some sight to the blind spot. Since 2009 the campaign, under Lundquist's stewardship, has authored beaver guidebooks, coordinated citizen-led surveys, and lobbied the state to account for beavers' benefits as well as their costs. Legitimizing beavers in the eyes of public agencies — a recalcitrant political network that Dolman terms the "egosystem" — is a full-time job. Lundquist and Dolman have espoused the rodents' virtues in lectures to Forest Service staff, convened workshops touting beaver restoration in the Sierra Nevada, and delivered more talks on the relationship between beavers and salmon than they can count. Thanks in large measure to their relentless crusading, beavers have moved from the radical fringe of California's restoration community to somewhere near its center. "The agricultural lobby has always wanted to lethally control beaver, and in the past there just hasn't been organizational will to move beyond those concerns," Lundquist told me. "Now we're seeing a lot of folks within the agency promoting beaver benefits." In validation of this growing esteem, the Salmon Restoration Federation — the very body whose members once showered Dolman with beaver innuendo — granted him their 2012 Golden Pipe award, an honor bestowed for innovation in fish recovery. The outsiders had become the cool kids.

That's encouraging, because you'd be hard-pressed to find a place that needs beavers more than California. The state's aquifers are catastrophically stressed: As of December 2016, twenty-one basins, nearly all in the Central Valley, were "critically overdrafted," meaning that far more groundwater was being withdrawn than replenished.[20] Although California has greeted past water crises by building bigger dams and longer pipes, the state no longer holds an infrastructural ace in the hole: Its best dam sites are developed, its biggest sources tapped, its proposed water transfer schemes prohibitively expensive. As Berkeley geologist Richard Walker told *California Magazine* in January 2018: "We're just rearranging the deck chairs on the *Titanic* instead of looking for icebergs."[21]

California may also have to soon make do with less — much less. By the time I visited in July 2017, the state was no longer sweating beneath the veil of its infamous five-year drought, but the dry spells will return. According to a 2015 study headed by scientists at Columbia University, the drought was exacerbated by hotter air, which can now hold eight and a half trillion gallons more moisture than it could a century ago. That evaporative power — think less water in rivers and reservoirs — will only grow with climate change, which, the researchers conclude, "has substantially increased the overall likelihood of extreme California droughts."[22]

Scientists generally divide our responses to climate change into two broad categories: *adaptation* and *mitigation*. The former strategies are reactive: As a turtle tucks her extremities into her shell to defend herself against danger, so we fortify our coastlines and plant drought-resistant crops. Mitigation strategies, on the other hand, are proactive — they're the measures we take to prevent the climate from changing too much to begin with, like taxing greenhouse gases, installing solar panels, and planting trees. Beavers, as Kent Woodruff preaches in the Methow, represent a promising adaptation strategy: By slowing, spreading, storing, and sinking meltwater and runoff, they can help us compensate, to some to-be-determined extent, for fast-vanishing glaciers and snowpack. They may also have a role to play in mitigation — in keeping carbon out of the atmosphere in the first place.

To be sure, beavers' climate impacts are anything but clear-cut. As ponds slow flows, they filter out carbon-rich organic matter — leaves, sticks, insect bodies, and just about anything else that lives and dies in the vicinity of streams. When that rich soup breaks down, it releases methane, a greenhouse gas thirty times more potent than carbon dioxide. In 2015 scientists found that the world's beaver ponds emit around eight hundred thousand metric tons of methane each year.[23] The study elicited hand-wringing from the climate-concerned media; one headline, perhaps tongue-in-cheek, warned about "The Latest Climate Change Threat: Beavers."[24]

Of course, in a global accounting of carbon emissions, beaver ponds wouldn't even register: Eight hundred thousand tons of methane might sound like a lot, but it's about 1 percent of cattle's contribution. And before we indict beavers for melting the Greenland ice sheet, consider this: Much of the organic material in their ponds remains stable. Just as forests suck carbon from the atmosphere and sequester it in wood, so beaver ponds lock it up in buried sediment. According to a 2013 study published by geomorphologist Ellen Wohl, active beaver complexes on twenty-seven streams in Rocky Mountain National Park once stored more than 2.6 million megagrams of carbon — by my calculations, the equivalent of thirty-seven thousand acres of average American forest.[25] Active colonies, Wohl found, store about three times more than derelict ones, and up to fourteen times more than dry grasslands. Forget trees: If you want to fight climate change, it's entirely possible you're better off planting beavers.

That's more or less what California is doing in Childs Meadow — 290 acres of damp snowpack-fed sponge, nestled high in the Sierra Nevada and framed by the spires of Lassen Volcanic National Park. Sierra high meadows are among California's most important ecosystems, the source of 60 percent of its water and home to more than half its biodiversity. They're also among its most imperiled: Around ninety thousand meadow acres face degradation by cattle, development, and climate change.[26] To Kate Lundquist and Brock Dolman, Occidental's beaver missionaries, the cure was as obvious as the ailment. In 2014 the duo brought Michael Pollock, the principal architect of Bridge Creek's Beaver Dam Analogues, to California to meet with The Nature Conservancy, which had purchased Childs Meadow seven years earlier to protect it from a golf course developer. By fortunate coincidence, beavers had already begun to prove their worth in Childs Meadow, colonizing its lower reaches and attracting endangered willow flycatchers and Cascades frogs. The Conservancy was so impressed by Pollock's vision and the beavers' handiwork that, in 2016, the group planted seven hundred willows, fenced the stream to deter cows, and pounded in posts to form six

The daemonelices scattered across the badlands initially perplexed scientists, but research eventually revealed them to be the burrows of *Palaeocastor* — an ancient, prairie-dog-like beaver. *Courtesy of the University of Nebraska.*

Although a beaver lodge appears impenetrable to the outside world, submerged entrances allow its inhabitants to come and go safe from predators.

A thirty-foot-long dam near Taos, New Mexico, transforms a straight, relatively featureless stream into a wonderland of pools and side channels.

Half-Tail Dale awaits a mate at the Methow Beaver Project's rodent love motel.

Skip Lisle does a spot of maintenance on one of his Beaver Deceivers, devices that protect roads from beaver-related flooding without having to kill the offending rodents.

Torre Stockard of the Methow Beaver Project sets a nonlethal Hancock trap, complete with willow bait, to capture and relocate a beaver accused of cutting down apple trees.

Catherine Means, Torre Stockard, and John Rohrer prepare to release a beaver into Washington's Methow Valley.

A dam spreads water in Scotland's Bamff Estate, where Paul and Louise Ramsay first introduced beavers in 2002.

Drew Reed of the Northern Rockies Trumpeter Swan Stewards relocates a beaver kit into a tributary of the Gros Ventre River in Wyoming.

Toothmarks scar a downed tree in Scotland, where beavers have recently been introduced after a four-century absence.

In Utah's Mill Creek, Mary O'Brien surveys a dam destroyed by anonymous human vandals.

A Beaver Dam Analogue, the joint creation of human and rodent labor, furnishes coho salmon habitat in California's Scott Valley.

Steelhead, including this juvenile captured in Oregon's Bridge Creek, are among the many species that benefit from beavers.

When humans build infrastructure, including railroads, in low-lying river valleys, conflicts with beavers often result.

Ruby Mountains, Nevada. "As animal life goes, that of the beaver stands among the best. . . . His lot is cast in poetic places." — Enos Mills, *In Beaver World*, 1913.

Photos captured in 1992, 2013, and 2017 illustrate how a combination of managed cattle grazing and beaver recovery restored the ecological and hydrological health of Susie Creek, a badly degraded stream in Elko County, Nevada. *Photos courtesy of Carol Evans.*

Beaver Dam Analogues. Kristen Wilson, a Nature Conservancy ecologist, told me researchers spotted three young Cascades frogs dwelling in the new human-built ponds the very next year.

With all due respect to frogs and flycatchers, though, the Childs Meadow project is driven by even grander concerns. California operates the world's fourth largest cap-and-trade program, an emissions reduction scheme that forces polluters to pay for the right to spew greenhouse gases. The state reinvests that revenue — $3.4 billion through 2017 — in a Greenhouse Gas Reduction Fund, a pot that pays for everything from low-carbon buses to reforestation to modernizing irrigation systems. It was that fund that footed the $30,000 bill for the Childs Meadow dam analogues, and provided another half-million dollars to study their impacts. Researchers plan to quantify carbon storage in the site's sediment and woody plants and, with luck, figure out how beavers can help us turn down the planet's thermostat.

In fairness, beaver ponds aren't the world's most reliable carbon-capture devices. "You can't expect beavers to remain in one place, whether it's because high flows blow out their dams or they run out of food or they're eaten by predators," Wilson pointed out to me. Beaver complexes are by their nature liminal, constantly metamorphosing from pond to spongy meadow to grassland and back again. Good luck to the modeler forced to calculate the climate contribution of such a protean landscape. If all goes well, the results of the Childs Meadow study, along with seven other Sierra meadow studies funded by the California Department of Fish and Wildlife, will eventually help the state understand carbon storage in restored meadows — although we'll likely have to wait until 2020 to know the results.

The department's willingness to fund beaver-based restoration offers further proof that, at long last, California's enmity toward beavers has begun to soften. Matt Meshriy, an environmental scientist with the Department of Fish and Wildlife's Upland Game Program, told me that the state has tried to let more beavers live: Rather than granting year-round trapping permits that allow irrigators to kill any unlucky beaver who ventures into their canals, the agency is issuing shorter-term permits that seek to target damage-causing beavers without excessive casualties. Although Meshriy said the state isn't actively reconsidering its strictures against relocation, he told me he could envision the prohibition loosening down the road. That could be especially consequential in the watery, low-lying Central Valley, the conflict hot spot where California's most prolific beaver populations collide with bountiful agriculture and fast-paced development.

"I do think there's an increasing willingness to live with beavers," Meshriy told me. "But that's running countercurrent to the increasing number of people living on the margins of the wild lands." California, like much of the country, is urbanizing and sprawling even as it's rewilding, a sure recipe for wildlife clashes. Coyotes stalk the streets of Sacramento, while drought-afflicted bears and bobcats wander into the Central Valley to slake their thirst in suburban backyards. In 2015, P-22, SoCal's most famous mountain lion, took refuge in a Los Angeles crawl space. Urban beaver conflicts might not be quite so dramatic, but they're ever more common. During my stay with Brock Dolman and Kate Lundquist, we visited the largest beaver complex in Sonoma County, a cluttered wetland whose bird- and frogsong is frequently drowned out by the roar of cars tearing around the track at the nearby Sonoma Raceway. "It's a pretty amazing statement about their adaptability that they can make a go of it here," Lundquist marveled as we surveyed the palatial lodge, scarcely a stone's throw from the highway.

And no one understands the resilience of urban beavers so well as the citizens of Martinez.

———

In the decade after the arrival of the Martinez beavers, Heidi Perryman tracked them with the devotion of a proud parent. She learned to recognize them by sight, pieced together elaborate genealogies, and followed the drama as if it were a daytime soap opera. The year 2010 was particularly full of heart-wrenching plot twists: The colony's matriarch died after she broke her upper incisors, forcing a two-year-old to assume the burden of caring for her younger siblings. The next year, the male vanished for a while before returning with a new mate. Perryman has watched twenty-five kits come of age in Alhambra Creek. "It's all very *As the Beaver Turns*," she told me.

In 2015 the Martinez drama entered its darkest season yet. All four kits born that year, along with one sub-adult, mysteriously died. State scientists necropsied the bodies and tested for contaminants, but found none. Whatever the cause of death, the parents, apparently having decided that Alhambra Creek was no place to raise children, skedaddled. Although beavers stopped by in 2016, they cleared out again well before August 5,

2017 — the tenth annual Martinez Beaver Festival, and the first ever held without its namesake animal in attendance.

The festival, held on a bright Saturday in the pocket park next to the Alhambra Creek bridge — hallowed ground for Believers — was a sweet affair: grown exponentially from its roots, still cute enough to charm. Although beavers were the honorees, it seemed to have evolved over the years into a general wildlife jamboree; Elise and I saw booths focused on the conservation of seals, coyotes, native pollinators, and birds of prey. Rusty Cohn, a photographer from Napa who spent years shadowing a beaver colony in a concrete-lined ditch, showed off his pictures in a bound book. Esteban Murschel, the Portland-based founder of a group called Beaver Ambassadors, distributed hand-drawn flipbooks. Perryman, looking overworked but happy, presided from a tent near the stage, trusty scrapbook by her side, dispensing nature tattoos and extolling the merits of her favorite rodent to the next generation of Beaver Believers.

Still, an air of loss hung over the proceedings. At festivals past, Jon had led tours along Alhambra Creek to visit the dams and lodges. Now the stream contained nothing but a green skein of algae and a couple of melancholic ducks. We tried to compensate for the absence of flesh-and-blood beavers by purchasing ersatz ones: At the silent auction, I placed the high bid for a beaver print and a beaver T-shirt, while Elise won a brass beaver bottle opener. I didn't get the sense we had much competition.

What the festival lacked in live beavers, it made up in Bucky Beaver, a character of Brock Dolman's invention. Dolman took the stage at one thirty sharp, decked out in a head-to-toe plush beaver costume endowed with the bulbous oblate head of a Teletubby. A beaver hand puppet perched on one claw like Mini-Me. To warm applause, he launched into a dizzying spoken-word monologue about all that beavers do to combat climate change, his monologue packed with Dolmanisms like "oil-ogarchy" and "plantcestors." We cheered on his spiel from beneath the shade of a live oak; in front of the stage, a volunteer ushered away a man attempting to light a cigar. "Right now, the planet is running a serious fossil *fool* fever," Dolman continued, undeterred. "You know what I'm saying — due to your collective craze of carbonaceous combustion creating cacophonous climate chaos." We whistled our approval. "Where we come from," he went on, "we say, Where there's a willow, there's a way-o." He blew the audience a kiss.

We found Dolman a few minutes later, back in street clothes, still flushed and damp. "Holy *hell* it was hot up there," he said as we high-fived.

"I assume you made that costume?" I said.

He looked pained. "Nah, didn't have time — I bought it." He leaned closer. "Dude, you do *not* want to search for beaver costumes online."

After the festival, we drove north along the Pacific for a few days, ending up in Olympic National Park for a backpacking trip. We didn't see any beaver sign in the park's old-growth rain forest, although our camp was invaded one night by a mountain beaver, *Aplodontia rufa* — an odd rodent, only distantly related to *Castor canadensis*, with the peaked face of a mole and the long, grotesque fingers of Mr. Burns. When we emerged from the backcountry, now eight days after the festival, I found, not to my surprise, that I'd received several emails from Heidi Perryman. The subject line on one message: "Are you sitting down?"

I clicked through to a YouTube video: the dark scrub of Alhambra Creek, its burbling the backing track to the swelling strings of Copland's "Appalachian Spring."[27] I saw a familiar-looking pile of peeled sticks, traversed by a curious skunk. And then the camera zoomed in on a dark mound of fur dabbling in the shallow creekbed, hands curled to mouth, sitting upright within expanding rings of concentric ripples. They were back.

—CHAPTER SEVEN—

Make the Desert Bloom

In November 1887 a twenty-nine-year-old politician named Theodore Roosevelt embarked on a five-week-long hunting trip into North Dakota's badlands, a lonesome range of alien buttes and endless skies. The badlands had for years been Roosevelt's sanctuary, the sacred terrain, said the New York native, upon which "the romance of my life began."[1] It was in the badlands that Roosevelt metamorphosed from an asthmatic urbanite to a strapping cowboy, where he learned to run cattle and shoot game, where the rough-and-tumble mythology of the future president was born and burnished. And it was to the badlands that Teddy retreated after his mother and his first wife, Alice, died within twelve hours of each other in February 1884, a day Roosevelt mourned in his journal by scrawling, "The light has gone out of my life."[2] North Dakota's spartan beauty and tough living — the oceanic plains, the trilling meadowlarks, the exhausting days and crisp nights — rekindled Teddy's joie de vivre.

By his 1887 hunting trip, though, Roosevelt's prairie paradise had gone dark. For three weeks Teddy — still skinny, but already sporting his trademark mustache, toothy grimace, and bespectacled squint — rode solo through the badlands, scanning for wildlife. He saw little. Elk, grizzly, and bison had vanished, and bighorn sheep and pronghorn grown scarce. Migratory birds had forsaken the Little Missouri Valley. The trigger-happy Roosevelt still bagged eight deer, four antelope, and a pair of sheep, but it was clear that his beloved landscape had slipped out of whack.

What had befallen the badlands? A wild mammal had been replaced by a domestic one. "There were so few beavers left," wrote Edmund Morris in *The Rise of Theodore Roosevelt*, ". . . that no new dams had been built, and the old ones were letting go; wherever this happened, ponds full of fish and wildfowl degenerated into dry, crack-bottomed creeks." The devolution was accelerated by the encroachment of cattle, which "eroded the rich carpet of grass that once held the soil in place. Sour deposits of cow-dung had poisoned the roots of wild-plum bushes, so that they no longer bore fruit; clear springs had been trampled into filthy sloughs; large tracts of land threatened to become desert."[3]

The sight pained Roosevelt, who, as a hunter and rancher himself, must have recognized his own culpability in razing the prairie. Back in New York, he convened a cabal of influential animal-lovers to "work for the preservation of the large game of this country," a group that later became the Boone and Crockett Club.[4] Roosevelt had always been a naturalist, but it was that mournful badlands trip that helped turn him into a conservationist. Seeing a landscape without its beavers can do that to a guy.

Teddy's seminal badlands experience provides a stark illustration of how badly cows and beavers tend to coexist. American ranchers own 120 million acres of private land and hold grazing leases on more than a quarter-billion public acres; where cattle roam, beavers are often scarce. That's partly because unchecked grazing, as Roosevelt discovered, demolishes streamside habitat, and partly because most ranchers consider the rodents irrigation-clogging, tree-felling, field-flooding menaces. "I like trees and I like cattle," Jonathan Knutson, a columnist at the trade journal *AgWeek*, griped in a 2016 screed titled "Nature's Engineers? Or Nature's Despoilers?" "In my personal experience, that leads — inevitably and unavoidably — to being none too fond of beavers."[5]

Hanes Holman is proof that there's nothing inevitable or unavoidable about detesting beavers. Holman, the manager of Elko Land and Livestock Company, is a thoughtful, amiable stockman fond of uttering provocative chestnuts like "We live in a world that is totally governed by time." I met Holman near a cattle yard in Elko County, Nevada — when I listened later to my recording of our conversation, I could barely make it out over the lowing of black Angus cows — not far from the PX Ranch, where he'd worked as a hand soon after graduating high school.

"I remember on the wall at PX, we would nail up beavertails, show how many we caught," he told me as we sat in a cottonwood's shade. "We had beaver everywhere. The irrigators were angry all the time. In the evening you went out and trapped beaver. It was just part of life, and you didn't know any better. My dad hated beaver, his dad hated beaver. That's what you did as a rancher — you fought beaver."

As Holman's understanding of riparian ecosystems evolved, so did his appreciation for aquatic rodents. Water is a Nevada rancher's most important asset, and any force capable of capturing and retaining it is a godsend. Now, Holman told me, "I'm probably one of the biggest advocates of the beaver." He's far from the species' only admirer in Elko County, where a series of promising grazing experiments suggest that beavers could become staunch agricultural allies. For that to happen, though, livestock producers — as tradition-bound a community as there is in the West — will have to reconsider their relationship with their aquatic nemesis.

"Half the industry is still on the other side of the fence, and you're never gonna get 'em all," Holman told me. "Just two months ago I had a pissed-off irrigator in my office who wanted to borrow a gun so he could get his beavers out of a culvert. Man has got a real bad way of trying to impose his will on nature."

———

At first blush it is hard to imagine a landscape less hospitable to beavers than the scrubby steppe that blankets Elko County. Elko sits squarely in America's coldest and northernmost desert, the Great Basin, a sweeping expanse of sagebrush, sagebrush, and more sagebrush that covers parts of California, Utah, Idaho, and Oregon, and nearly the entirety of Nevada. Once, the Great Basin was wetted by marshes and the forebear of the Great Salt Lake; today it is a vast sump trapped between the rain shadows cast by the Sierra Nevada and the Wasatch Range. Not a single drop of water that falls within the Great Basin reaches the ocean, collecting instead in saline lakes, draining into the soil, or evaporating into parched air.

Although the Paiute thrived here for millennia, white folks found the basin impossibly forbidding. One writer called it "God-forsaken country that never was designed to be the habitation of a Christian or civilized man."[6] In 1890 an Elko cattle company lost 98 percent of its livestock to a single brutal winter.[7] Today it's commonly known as the Big Empty. The Great Basin has its charms, but even its partisans admit it's an acquired taste.

Among the region's fans is a fish biologist named Carol Evans, who, during the final week of spring, drove me around Elko's stark rangelands to search, improbably, for beavers. We took her truck, a rust-colored Ford whose undercarriage thumped hard against humps in the dirt road. Jackrabbits wove out of the sagebrush, flirting dangerously with the tires. Though it was only 8 AM, the sun already hung white and oppressive. There was hardly a tree in sight — just gray sage and yellow grass, rising and falling like wind-tossed crests.

Evans, whose sun-bleached hair and permanent tan betrayed a lifetime spent wandering the desert, asked me what I thought of the scenery. I told her I was impressed by its dryness. She laughed. "This is about as green as it gets in Nevada."

Evans, a Reno native, first came to Elko County to work for the US Forest Service in 1977, eventually ending up, in 1988, at the federal Bureau of Land Management. She never left. Her career is a testament to the value of long-term, place-based knowledge, to the power of observing a landscape over decades until its every blade of grass has become an acquaintance. "Whatever I've learned here, it's only because I've stayed for thirty years," Evans told me. "That doesn't happen in the agencies. The turnover is high." Although she retired from the bureau in 2016, she still works as a sort of scientist emeritus, consulting with conservation groups and ranchers on their never-ending management problems. Touchingly, she told me she longs to write a book called *Stream Stories*, biographies of the creeks she's spent her lifetime monitoring and rehabilitating. She has the gentle manner of a painting instructor and, rare for a federal employee in this corner of Nevada, the total respect of cattlemen. "When you can work with a person who has things figured out like Ms. Carol," one rancher told me, "you can get some stuff done."

We passed through a gate, just a few strands of barbed wire looped around posts, and parked above a steep ravine. Below us stretched an oasis: the sparkling thread of Susie Creek, its single sinuous channel winding through a valley the width of a football field. The water ran so clear that, even from twenty feet up, I could see crayfish creeping along its sandy bed. The floodplain was lush, at least for Elko, a green expanse of sedges and rushes. Killdeer flashed through the valley on black-and-white wings. Evans snapped a few pictures with her Pentax DSLR. To appreciate just how remarkable this lovely ribbon of green habitat is today, she told me, you have to understand what it once was. If *Stream Stories* ever gets written, I'm confident that Susie Creek will be the best chapter.

When Evans first encountered Susie Creek in the late 1980s, she found the stream in an abject state. The watershed had endured heavy grazing by cattle, sheep, and horses since 1875. The banks, denuded of willows and sedges by thousands of hungry mouths, crumbled further with each flood. Over the decades the stream, like Bridge Creek and so many others, chewed deeper into the earth, at last entrenching itself within a gully. The water table plummeted; temperatures rose; lush meadows dried out and were overrun with sagebrush and rabbitbrush. By the time cattlemen realized they'd erred, it was too late. In his book *A Long Dust on the Desert*, cowboy Ed Hanks described a depressing conversation with an old-timer named Fent Fulkerson:

> *Fent used to tell me how almost all the creeks on the northside of the Humboldt River ran almost on top of the ground, when he first came to the country. These creeks, Maggie Creek, Susie Creek, Rock Creek, and Antelope Creek, to name a few, overflowed and made an abundance of feed for livestock.... These creeks today have washed down to bed rock and it isn't easy to get the water out, so there isn't as much feed.[8]*

In Nevada, and throughout the West, most ranchers graze their livestock on federal land, paying the government rent for the privilege. For decades, the degradation of those lands seemed inevitable. Healthy riparian areas provide ten to fifteen times more forage than the surrounding uplands, and in many high deserts hold the only water for miles, luring cattle into streambeds to graze, drink, and wallow.[9] The attraction grows stronger in summer, when feed and water grow scarcer. Although grazing during "hot season" — generally late June to late September in the Great Basin — inflicts the most concentrated damage upon riparian plants, few objected to the practice. Of course cows preferred to hang out along streams — why fight it? If that meant overgrazing, erosion, and impaired water quality — all those cow patties had to go *somewhere*, after all — well, that was just life on the range. "Riparian areas were considered sacrifice zones," Evans said. "People just concentrated on managing the uplands."

Evans, however, wasn't willing to sacrifice Susie Creek. She was, after all, a fish biologist, tasked with recovering a threatened species called the Lahontan cutthroat trout. Lahontan cutthroat are North America's largest trout — early settlers hauled sixty-pound specimens out of Pyramid Lake — and among its most beautiful, ornamented with pink-gold flanks and black freckles. Unfortunately, Nevada has eliminated its official state fish from more than 90 percent of its streams and lakes, including Susie Creek, where unshaded banks and hot water drove out Lahontan cutthroat decades ago.[10]

In a photo Evans took in 1989, Susie resembles a Martian canal: It trickles across a stony, barren floodplain locked deep within eroding canyon walls, undecorated by so much as a blade of grass. It's no place for a trout, a mule deer, or even a cow — a stream not fit for man or beast. Susie Creek's story, however, had only begun.

———

If you had to choose the least likely place in the West to reform cattle grazing, northeast Nevada might be it. Around 85 percent of Nevada is managed by federal agencies — the highest percentage of any state, and a source of resentment for many rural conservatives.[11] For nearly a century Nevada's ranchers, miners, and loggers have bridled against restrictions on land and water use, sparring with the Bureau of Land Management and the Forest Service over everything from grazing permits to road closures. In the 1970s Elko County was a staging ground for the Sagebrush Rebellion, a movement that sought to seize control of federal lands and turn them over to states and counties. Although the rebellion fizzled, protests flared throughout the 1990s and early 2000s. Some federal employees, exhausted by harassment and fearful for their lives, skipped town. "Officials at all levels of government in Nevada participate in this irresponsible fed bashing," one Forest Service supervisor griped on her way out the door.[12]

Given that vitriolic history, you could be forgiven for assuming that the Bureau of Land Management's plans to reform riparian grazing would fall on deaf ears. And at first Jon Griggs was indeed skeptical. In 1991 Griggs, a Nevada native with windburnt cheeks, a dry sense of humor, and a basso voice, came to work as a cowboy on Maggie Creek Ranch, a two-hundred-thousand-acre quilt of public and private land that enfolds the Susie Creek watershed. (Yes, Maggie Creek Ranch grazes cattle in the Susie Creek watershed; it's confusing, I know.) To Griggs, the moonscape looked normal. "If anything, I thought that was a pretty dang good July because there was water in the creek at all," he told me at the ranch's headquarters, a light-filled office conspicuously decorated with twenty-five years of environmental trophies and plaques. "I took some convincing. I didn't know the potential of it."

Carol Evans's prescription was simple: prevent cows from overgrazing Susie Creek's meadows during hot season, when plants were most vulnerable. The ranch's then-manager, Wayne Fahsholtz, cautiously consented to install miles of additional fencing along the stream, partitioning the allotment into several enclosed pastures. No longer were cattle permitted to wander willy-nilly any time of year; now Griggs and the other cowboys could surgically rotate their black Angus from pasture to pasture, keeping herds out of delicate riparian areas during summer. The strategy didn't reduce grazing pressure so much as it modulated when and where cattle could roam. The Bureau suggested a similar approach to ranchers on neighboring streams, who also agreed to play along. Then they sat back and waited.

They didn't have to wait long. By 1994 streamside plants, protected each summer from rumens and hooves, had recolonized hundreds of acres. Susie Creek, Maggie Creek, Dixie Creek — one degraded gully after another greened up with willows, rushes, and sedges. No longer did floods tear through rocky channels, gouging out beds; instead, revived vegetation slowed down the water, settling out sediment and rebuilding the floodplain. Susie Creek and its sister waterways stopped *degrading* and began *aggrading* — climbing toward their long-disconnected floodplains like plants seeking sunlight. By the time Evans snapped her annual Susie Creek photograph in 1999, willows filled the frame. And where there are willows, beavers are not far behind.

When the first rodent pioneers arrived in Susie Creek in 2003, Jon Griggs, who'd become ranch manager in 1998, felt conflicted once more. "As stockmen, we want so much to control the things we can control, because there's so much we can't," he told me. "We can't control the weather. We can't control our markets. We can't control government regulation. But when beavers show up, we can control that." Control, of course, means kill. When he mentioned the intruders to Evans, though, she advised him, again, to wait and see. The cattleman had come to trust the fish biologist, so he let the beavers live. Almost immediately, he was glad he did.

If managed grazing changed Susie Creek, the arrival of beavers transformed it. Narrow channels became sprawling cattail marshes. Water-loathing sagebrush ceded to native sedges. The slender green string twisting through the desert broadened into a thick band. When scientists at the nonprofit Trout Unlimited compared aerial photos from 1991 to pictures from 2013, they found that Susie Creek contained twenty more acres of open water, a hundred more acres of riparian vegetation, and three additional miles of wetted channel. Satellite imagery suggested that photosynthesis, a rough measure of plant productivity, had increased by 32 percent since the late 1800s. Water tables rose as much as two feet; floodplains rebuilt, too. Once-beaverless Susie Creek now churned through a labyrinth of 139 dams.[13] A wasteland had become a paradise.

For Griggs, the stream's evolution proved salvational. The beaver-irrigated meadows along Susie Creek became his most vital pastures, a haven for calves during the vulnerable days after their separation from their mothers. "The worst thing for cattle is stress," Griggs said. "So you think about calves going into that green lush feed when everything else around it is dormant, staying there for a couple of days at a time when they're facing probably the most stress of their lives – that's a pretty big bump they're getting."

But it wasn't until 2012 that Griggs's rodent partners revealed their true worth. That year northeast Nevada was gripped by the worst drought he'd ever seen, an unrelenting dry spell that lingered into the next year and the next, all the way into 2015. Many ranchers had to pull their cows off the range or swallow the crippling expense of trucking water to their livestock. Griggs, on the other hand, managed to avoid hauling water – thanks to beavers, who kept Susie Creek brimming through the driest months.

"Jon Griggs is a pretty big guy in the community," Hanes Holman, whose ranch profited similarly from beavers, told me. "Every rancher in the state went, 'Our creeks are dry, but what the heck is Griggs doing over there? He's got more water than he knows what to do with.' Ranchers are economically driven – they have to see the value in these projects. You wouldn't sell one rancher on beavers if you couldn't prove that increased water came with it."

Cows aren't the only desert dwellers who benefit from beavers. Disconcerting climate projections suggest that warming will force Lahontan cutthroat trout from Nevada's low-elevation streams before long. Just as beaver dams made Bridge Creek better suited for steelhead, Carol Evans hopes the rodents will someday convert Susie Creek into a refuge capable of supporting cutthroat, climate change be damned.

"This is still an unfolding story, and we've learned so much about how these systems function," Evans said as we traipsed across the Susie Creek floodplain, skirting the gnarled skeletons of sagebrush killed by rising water tables — striking visual evidence of the valley's rehydration. "We're getting a glimpse into something here that we hadn't seen in our lifetimes."

Although the Elko acolytes may be the livestock community's most enthusiastic beaver proponents, they're far from the only ranchers reexamining their rodent relationships. Alan Newport, the Oklahoma-based editor of *Beef Producer* magazine, captured the industry's fitful evolution in a 2017 column detailing how beavers could forestall the fatal gullying of some rangeland streams. "I understand that beavers are a pain in the neck, but so is erosion and droughty land," Newport wrote. "So here I stand, saying kind things about one of the most hated creatures in the world of agriculture."[14]

In fact, a minority of ranchers have always embraced beavers, at times going to extraordinary lengths — and no doubt risking their neighbors' opprobrium — to avoid killing the critters. Joseph Grinnell, the California biologist who misdiagnosed the beaver's native range, reported in 1937 that alfalfa growers in the Merced River Valley so championed beavers' water-saving abilities that they refused to slaughter the animals, even when they dammed irrigation ditches. One stockman, incredibly, opted instead for corporal punishment. "These trapped beavers were soundly spanked by the rancher, who used a good stout board for this purpose," Grinnell wrote. Clubbing animals with planks, it turns out, doesn't make for a particularly effective management strategy. "After they had been spanked and turned loose they stayed away for a while," Grinnell added, "but a few weeks later some of the same beavers, identified by the trap marks on their toes, were again caught in this canal."[15]

The most compelling tale of beavers' drought-fighting abilities comes from Glynnis Hood, an ecologist who studied castors in Elk Island National Park, a preserve outside Edmonton where the rodents were trapped out and later reintroduced. After drought struck Alberta in 2002, Hood wrote in her book, *The Beaver Manifesto*, "Beavers excavated channels to maintain and direct water where they needed it most — at the entrances to their lodges and along access routes to their favoured foraging areas."[16] When Hood compared aerial photos of Elk Island snapped back in 1950, before beavers reclaimed the park, to more recent pictures, she found that ponds contained 61 percent more open water in 2002, when the four-legged hydrologists were on hand to capture life-giving liquid.[17]

"Cattle ranchers were one of the agricultural groups hit hardest by the drought in the region," Hood added, "and it was common for them to request access to a neighbour's property if there was an active beaver lodge on it."[18]

Beavers not only benefit agriculture, they also protect the public from some of food production's worst side effects. Every year, America's farmers collectively add around twenty million tons of fertilizers to their fields — nearly 140 pounds per acre. Rain sweeps much of that excess nitrogen and phosphorus into rivers and, eventually, to lakes and seas. (Suburban lawns, septic tanks, and even cars add nitrogen to watersheds, too.) The nutrient stew fertilizes algal blooms that die and decompose, devouring dissolved oxygen and giving rise to "dead zones," lifeless expanses of anoxic water that drive away fish and kill stationary bottom dwellers like clams and mussels. Global oceans are afflicted by nearly a hundred thousand square miles of dead zones; the one that forms annually in the Gulf of Mexico, where the Mississippi River deposits pollution gathered from Pennsylvania to Montana, could swamp the state of New Jersey.[19]

One solution to this marine crisis is wetlands, which, like kidneys, filter out suspended nutrients and other pollutants long before they reach the sea. Restoring marshes, swamps, bogs, and fens ranks among our most effective strategies for treating farm runoff. A conservation program in Iowa, where 95 percent of wetlands were drained by 1930, found that constructed wetlands extracted up to 70 percent of nitrates — nitrogen-based fertilizer additives — and more than 90 percent of herbicides from agricultural drainage.[20]

If constructing wetlands is good, letting beavers do it for free is even better. In one 2000 study on Maryland's coastal plain, researchers found that a beaver pond slashed the discharge of total nitrogen by 18 percent, phosphorus by 21 percent, and total suspended solids — waterborne particles that are classified as a pollutant by the Clean Water Act — by 27 percent.[21] In Quebec, Bob Naiman found that beaver ponds stored sediments containing around *one thousand times* more nitrogen than unaltered riffles.[22] At Lake Tahoe, where locals preach the importance of water quality with religious fervor, demolishing beaver dams on the lake's tributaries more than doubled the phosphorus flowing into the once-crystalline body of water.[23] All the KEEP TAHOE BLUE stickers that adorn bumpers in Northern California might as well read KEEP TAHOE'S BEAVERS.

And it's not just that beaver ponds capture and store nutrients — thanks to one 2015 paper, we know they can even change pollutants' physical state. In the course of their study, researchers from the University of Rhode Island jammed metal tubes into the squelchy bottoms of three rodent-built ponds to collect soil cores. Then they applied nitrate to the samples and tracked the molecule's fate. Bacteria living in the sediment broke down the nitrate, performing a microbial alchemy, called *denitrification*, that effectively purged the pollutant from the water by converting it to nitrogen gas. The researchers calculated that beaver ponds, and the microscopic wizards living in their soils, could remove up to 45 percent of nitrates from southern New England's rural watersheds, preventing dead zones from forming in its estuaries.[24]

"We have a species whose population crashed from widespread trapping 150 years ago," Arthur Gold, the hydrologist who led the research, said when the study was published. "With their return, they help solve one of the major problems of the 21st century. I don't want to minimize that. We have to remember that those ponds wouldn't be there without the beavers."[25]

———

Susie Creek is a powerful testament to the notion that beavers can thrive alongside properly managed cattle. But Jon Griggs and Hanes Holman, for all their benevolent influence, remain outliers in the world of agriculture. For a window into how beavers and livestock typically coexist — or fail to — look six hundred miles southeast of Elko County, to a landscape nearly as dry as northern Nevada: New Mexico.

The Land of Enchantment holds a prominent place in beaver history. In the 1820s, while Mexico still controlled the region, scores of fortune-seeking Americans trekked southwest along the Santa Fe Trail to trade and trap. While the mountain men stripped the Northern Rockies bare, the

Taos Trappers hammered the Sangre de Cristos. In 1824 alone, estimated the author William deBuys, "the year's gross sales represented the skins of about two thousand animals — an enormous slaughter." In 1838 officials prohibited trapping beavers and otters in the Rio Grande, one of the West's first conservation laws, although deBuys pointed out that the rules were "probably unenforceable."[26]

Like trapping, grazing came early to New Mexico. While cows didn't arrive in the West at large until after the Civil War, Spanish *criollo* cattle began creeping into the Upper Rio Grande as early as the sixteenth century.[27] The longevity of cattle's influence may help explain why New Mexico's beavers remain relatively scarce today. In areas with healthy populations elsewhere, ecologists generally observe between one and three beaver colonies sharing a mile of stream. But when Brian Small, then a master's student at the University of New Mexico, scoured the state's public lands for beavers, he found but one colony every *fifty miles*. Wherever the Forest Service permitted cattle to graze, streamside plants — and the aquatic rodents that depend on them — remained scarce. "Until livestock grazing is managed in a way that produces adequate amounts of willow growth," Small warned in a 2016 study, "the restoration of dam-building beaver is not likely to be possible in many areas."[28]

One perfect fall afternoon, a day when every aspen and cottonwood in New Mexico seemed to have burst into golden glory in unison, Jim Matison took me to the Santa Fe National Forest to show me what unchecked grazing can do to beavers. Matison serves as restoration director for WildEarth Guardians, a scrappy Santa Fe-based conservation group; among the streams he's reviving is one known, almost too perfectly, as Rio de las Vacas, or River of the Cows. As we passed a private inholding tucked away in National Forest land, I saw how the stream earned its moniker: A dozen brown-and-white cattle lounged around a shallow, entrenched, barren gulley. Nary an alder poked above the table-flat floodplain. Matison was wearing sunglasses, but I could almost see his eyes roll. "That's what we're trying to change," he said.

We drove on, and soon came to Matison's pride and joy: a three-mile stretch of stream bordered by towering mesh fences, a series of exclosures that kept cows — as well as wild grazers like elk and deer — out of the River of the Cows. The project had been funded primarily by a pollution control grant from the Environmental Protection Agency; the "pollutant," in this case, was the unnaturally warm water, often lethal to trout, that dribbled through this degraded channel. By thwarting grazing and planting willow and alder, Matison hoped to encourage the return of beavers and, with them, the rebuilding of a channel incised by centuries of grazing. The project's ultimate goal was to reconnect the river with its floodplain, to foster the kind of supercooling surface-to-groundwater recharge and exchange that Nick Weber documented in Oregon's Bridge Creek.

WildEarth Guardians had only completed the fences the previous year — the group hired local Navajo and Jemez Pueblo teenagers to construct the exclosures — but the newly cow-free stream had already lured a nearby beaver colony. A series of small, slightly ramshackle dams, which the emigrating rodents had built from rocks and planted willow stems, accented the creek. In some places, Matison pointed out, beavers had actually tunneled into the bank to tear out wire-wrapped willows by their roots. Although I'd heard other restorationists complain about beavers interfering with their plantings, Matison seemed happy to feed his tenants. "I'll say in my grants, we're gonna plant thirty or forty thousand willows," Matison told me. "Funders will look at me like I'm crazy. 'Why do you need all that? That's overkill.' But it's not. Not when you're trying to support beavers."

Although beavers have returned to Rio de las Vacas, the overwhelming majority of stream miles in New Mexico remain destitute — and, barring a policy change, they'll stay that way. Even if exclosures grew back habitat, it could take the rodents decades to infiltrate many streams. And in New Mexico, transplanting beavers isn't really an option: "The relocation rules are horrific," Matison told me. The state's Game and Fish Department has to obtain the consent of any landowners within five miles of the release point, and, in such a cow-dominated land, that approval is rarely forthcoming. Never has the state granted one of WildEarth Guardians' relocation requests.[29] "It's great that there are some ranchers who understand the benefit of beavers," John Horning, WildEarth Guardians' executive director, told me before my tour with Matison. "But for every one of those guys, there are ten who still vilify them."

Truthfully, it's hard to blame them. A primary challenge of beaver-based restoration is, I think, the misalignment of its costs and benefits. By this, I mean that while the benefits — fish and wildlife habitat, cleaner water, carbon storage — tend to accrue to the general public, *private* landowners are the ones who find their cottonwoods felled, their irrigation ditches plugged, and their alfalfa underwater. Persuading ranchers to endorse beavers thus requires convincing them that it's in their best interests — and dispelling generations of ingrained anti-castorid bias. In Elko, it was water storage and forage production that convinced Jon Griggs to stop worrying and love the beaver. On ranches elsewhere in the American West, the incentive comes in the form of a chunky, spike-tailed, ground-nesting bird.

From a land management perspective, the greater sage grouse might be the most important bird in America. It certainly has a claim at the title of strangest. To woo paramours, male grouse congregate en masse at mating grounds called *leks*, inflate bulbous yellow chest pouches, and jostle their balloon-like breasts to produce a corky *pop*. Sage grouse leks once dotted the sagebrush sea, an expanse of high desert that spans 270,000 square miles spread across eleven western states. As the Big Empty has become the Big Occupied, a litany of threats — energy and residential development, transmission lines and roads, the incursion of invasive cheatgrass — has fragmented the West's unbroken expanse of grouse habitat into a perilous checkerboard. Perhaps sixteen million sage grouse once strutted their treeless, wind-blasted kingdom; these days the bird's population hovers around four hundred thousand.

Although sage grouse are impeccably adapted to their desert domain, they spend their summers chasing water. During spring melt, the steppe bursts into bloom, and downy chicks gorge on clover, phlox, biscuitroot, and a smorgasbord of other wildflowers, as well as the insects those plants attract. As the days get hotter and vegetation withers, grouse go on the march, following the receding green wave to the few wetlands and streams that still harbor palatable plants — emerald islands in a sagebrush sea. Beaver trapping, historic overgrazing, and other pressures have degraded those islands; in many corners of the sage grouse's range, the green archipelago covers less than 2 percent of the landscape. "We've essentially lowered the land's carrying capacity by providing fewer resources that are more widely spread apart," Jeremy Maestas, an ecologist with the US Department of Agriculture's Natural Resources Conservation Service (NRCS), lamented to me. "A hen and her little chicks might have to walk fifteen miles, and they all have to pile into one spot where predators can focus on eating grouse all summer."

The vagaries of sage grouse habitat aren't merely academic. In 2010 the US Fish and Wildlife Service ruled that grouse were "warranted but precluded" from Endangered Species Act protection — deciding, in effect, that although the grouse was in trouble, the agency's limited conservation resources had to be allocated to even more precarious creatures. The government resolved to revisit the grouse's plight in 2015, hurling western industries and agencies into panic. The land-use restrictions that would come with an endangered listing, many feared, would be catastrophic for sectors from ranching to wind energy. *The Washington Post* claimed that state governments "[stood] to lose billions of dollars in tax revenue and economic activity."[30]

No sooner had the ink dried on the 2010 decision than the NRCS launched the Sage Grouse Initiative, a preemptive strike against a potential endangered listing. The initiative aimed to help ranchers and agencies *voluntarily* save grouse through proactive conservation measures, like controlling invasive plants, protecting land from development under conservation easements, and removing fences that clothesline flying birds. The "largest conservation effort in history," as one scientist put it,[31] convinced the powers that be: In September 2015 Interior Secretary Sally Jewell announced that grouse no longer needed Endangered Species Act protections — contingent, of course, on all those voluntary conservation actions being carried out.[32]

While most grouse habitat restoration occurs in the sagebrush uplands, conserving the bird unavoidably requires restoring wet meadows — and it's damn near impossible to do that without roping in beavers. Jeremy Maestas, the Sage Grouse Initiative's ecosystem specialist, lives in the high desert of central Oregon, not far from Bridge Creek, where biologists jacked up baby salmon survival by building artificial beaver dams. In 2015 Maestas caught wind of the Beaver Dam Analogue installations and toured the site. "That project had nothing to do with sage grouse," he told me. "But what I saw was more water getting onto the floodplain, soaking into the banks, raising an incised channel, growing more riparian vegetation that stayed productive throughout the summer. Those are the things that grouse need, and those are the things that livestock producers want." What's good for the bird, as they say, is good for the herd.

The daunting geography of sage grouse country provided another inducement. "We've got to find techniques that are lower cost, that more people can engage in, that allow a system to take over and fix itself," Maestas told me. Since 2015 Utah, Idaho, Oregon, Washington, Montana, and Wyoming have all hosted Beaver Dam Analogue test sites and trainings, and Maestas conducts workshops and webinars on the subject. Although the federal government manages around half the American West's acreage, 80 percent of the region's wet habitats fall on private land — after all, you can't found a homestead without settling near water.[33] Restoring streams, wetlands, and damp meadows in sage grouse territory therefore requires convincing farmers and ranchers to embrace their sharp-incisored nemesis.

"Some folks are at first not willing to talk about beaver," Maestas said. "But if I told them that we had a way to simulate what beaver *do*, they'd probably work with me. Pretty soon they say, 'I really like how these structures keep water on the land and produce green groceries for my cows, but I don't want to maintain them myself every year.' And that's when we can have the next conversation: Well, what about getting beaver back here and turning this into a self-sustaining system?" Sage grouse hold the door open, and *Castor canadensis* waddles on through.

———

One rancher who requires no convincing about beavers' merits is James Rogers, the jocular, pink-faced manager of the Winecup Gamble Ranch. At nearly a million acres, the Winecup Gamble, a mountainous Nevada spread hard up against the Utah border, puts Rhode Island to shame. A slick promotional video touts the ranch's world-class elk hunting and trout fishing. When I visited in 2017, the Winecup Gamble was owned by Paul Fireman, the executive behind the ascent of Reebok; its past owners include a Nevada governor and Jimmy Stewart.

Posterity failed to record how Jimmy Stewart felt about beavers, but it's safe to say his ranch manager didn't look kindly upon them. When James Rogers first assumed control of the Winecup Gamble in 2010, he found that his cowboys shot the rodents on sight, as reflexive as breathing. Rogers's own father, a rancher back in Wyoming, was no great beaver admirer — he dynamited their dams or ripped the structures out with a backhoe — but somewhere along the line James had converted to Beaver Belief. "You just see glimpses of the benefits, if you have enough of an open mind," he told me as we bumped through one of his pastures on a June afternoon, his truck enveloped by a cloud of dust. In the world of livestock production, where conventional wisdom and machismo prevail, Rogers's gee-whiz humility marked him as anomalous. "I'm always looking for ways to keep water here, and the beaver do it for free. They don't do it perfectly, but who am I to think that I can do it better?"

Rogers pulled over and led me on foot downhill, into a valley lushly furred in bunchgrass. Thousand Springs Creek, so named for its countless seeps and springs, glugged along, impounded every hundred yards or so by another willow beaver dam. "This valley is probably producing one thousand pounds of grass per acre, and the sagebrush we just walked through is only a hundred," Rogers said, running a stalk through his fingers. "From a rancher's perspective, beavers increase production tenfold." Ever since Rogers declared an armistice in the Winecup Gamble's war against beavers, he told me, the creek's castorid population had been expanding its range; lately beavers had ventured down from the slopes and pushed toward ranch headquarters. Rather than shooting road-threatening beavers, though, he'd hired a conservation group called the Seventh Generation Institute to lead a flow device installation workshop.

"I want everybody to buy in — I want people to feel as passionate about 'em as I do," he told me when we stopped to admire a particularly majestic dam. "When a new guy comes, I don't want to be the guy to tell him we don't shoot beaver on this ranch. I want my other employees to say, man, you gotta check out what these beaver are doing, they're awesome — and by the way, we don't kill 'em. I want us to all be on the same page."

In the winter of 2017, Rogers's relationship with beavers took another leap forward — thanks not to drought, but to flood. Ferocious storms swept Nevada and California, washing out roads, smothering highways in mud, and killing two people. On February 8 disaster struck the Winecup Gamble. Twenty-One Mile Dam, a forty-seven-foot-tall earthen structure that supplied irrigation water to the ranch, caved beneath its swollen reservoir. Floodwaters scoured Route 233, rerouted trains, and forced Elko County to declare a state of emergency during the "dangerous and life-threatening" deluge, though in the end no one, fortunately, was injured.[34]

If you missed the Twenty-One Mile Dam story, it's because that breach was overshadowed by an even more terrifying near-miss. As the Winecup Gamble's dam was crumbling, the Oroville Dam, a 770-foot earthfill embankment on California's Feather River, was buckling, too. The same storms that hammered Nevada also bloated Lake Oroville to perilous levels, forcing engineers to dump water down the dam's concrete spillway – a reasonable idea, until a crack in the spillway grew into a 250-foot crater. When administrators closed that channel, rising waters began cascading over the lip of the tallest dam in the United States, eroding hillsides and threatening to undercut the structure. California officials, with visions of a lethal thirty-foot wall of water dancing in their heads, evacuated nearly two hundred thousand downstream residents.[35] The dam ultimately survived, though the close call shook engineers and water managers to the core.

In retrospect the most remarkable thing about the Oroville near-catastrophe was not that the dam almost failed but that disasters don't strike more frequently. America, as one documentary put it, is Dam Nation: Our rivers are interrupted by more than ninety thousand walls, the most famed of which – the Hoover, the Glen Canyon, the Bonneville – rank among the largest structures on earth. Many dams, by controlling floods, generating kilowatts, and irrigating farmland, made possible large-scale colonization of the arid West; many, however, are aging blights that were follies from the moment their foundations were poured. As Marc Reisner made clear in *Cadillac Desert*, most major dams began their lives as pork-barrel boondoggles that spent billions of public dollars to achieve mere millions in private irrigation benefits. With the dam-crazy Bureau of Reclamation at the helm, Reisner wrote, "the American West quietly became the first and most durable example of the modern welfare state."[36]

The last major dam failure in the West occurred when Idaho's Teton Dam disintegrated in 1976, killing eleven people and thousands of cattle. Collapses may be mercifully rare, but many dams still pose cataclysmic risks. In its 2017 Infrastructure Report Card, the American Society of Civil Engineers classified more than fifteen thousand dams as "high-hazard." Rehabilitating those dangerous dams alone, the report found, would cost twenty-two billion dollars. Repairing our entire decrepit fleet of dams, whose age averages fifty-six years, would run us more than sixty billion.[37]

Even when they're not crumbling, dams wreak havoc – there's a reason John McPhee placed them at conservationists' "absolute epicenter of Hell on earth."[38] They gut ecosystems: In the Colorado River, less a river at this point than a chain of reservoirs, fishes like the pikeminnow and the razorback sucker have been nearly wiped out by changing water temperatures, altered sedimentation patterns, and invasive species. Although dams are touted as sources of clean power, methane outputs from their reservoirs account for more than 1 percent of global emissions.[39] And, finally, they fill in with silt, rendering costly public investments inoperable. More than half the Bureau of Reclamation's dams are over sixty years old, approaching the end of their "sediment design life."[40] When beaver ponds fill in with sediment, they become meadows; when vast reservoirs do it, they become bowls of shallow, turbid soup, incapable of spinning turbines or sustaining aquatic life.

It is worth pondering, for a moment, why the first wave of farmers who invaded the West found so much land they needed to "reclaim." We associate the Southwest with hot sand and red rock, stunted mesquite and prickly pear. Once, though, it was a much lusher place — thanks in part to beavers. James Pattie, a trapper who wandered southwestern New Mexico in the 1820s, described a landscape cratered by ponds and wetlands; attracted one day to a small lake by the honks of ducks and geese, he "remarked what gave me much more satisfaction, that is to say, three beaver lodges."[41] On one river, Pattie claimed to have taken "the very considerable number of 250 beavers,"[42] no doubt destroying in the process hundreds of dams and the watery *cienegas* — spongy, alkaline wetlands endemic to the Southwest — they helped create. Having depleted our landscapes' ability to store water, we eventually felt compelled to imitate our rodent rivals by building thousands of concrete boondoggles. Our dams were partly designed to solve a problem trappers created a century earlier.

At this point, you are perhaps shaking your head at the notion that a passel of rodents, however industrious, could match the capacity of, say, Lake Mead, the country's largest reservoir, which waters everything from the Las Vegas Strip to lettuce heads in California's Imperial Valley. "I'd be cautious about calling beavers even a partial solution for some of these gigantic storage and delivery projects," Jerry Meral, former water commissioner for California's Marin County, told me. He's an ardent beaver proponent, but he's also grateful for human infrastructure. "California has built seventeen hundred dams, including some of the largest ones in the world, to store water in winter and move it so it could be used for irrigation season in summer. There's nothing a beaver could do to match that."

Beavers would certainly be hard-pressed to replicate the precision of the West's extraordinary water-delivery infrastructure. For all the risks that come with centralized water storage, there are also many advantages, notably the convenience of sticking a bunch of straws into a single milkshake. But it's not far-fetched to imagine that a revitalized southwestern beaver population could at least equal Mead's raw storage capacity. Although the reservoir can hold more than thirty-one million acre-feet of water — a unit equivalent to the volume you'd need to submerge an acre of land beneath a foot of water — years of drought and withdrawals have dropped its surface more than 140 feet, leaving chalky bathtub rings circling its walls.[43] As of this writing, Lake Mead contains just 38 percent of its capacity — around eleven million acre-feet.

Could beavers top that? Variations in geology — for instance, differences in the porosity of underlying rock — make it difficult to compare rodent-driven water storage across regions. But one study conducted by the Lands Council in arid eastern Washington estimated that beaver dams had the potential to store between 17.5 and 35 acre-feet of combined surface water and groundwater apiece.[44] If you permit me the liberty of applying those results to the Southwest, it would take roughly 320,000 to 620,000 new beaver dams — the work, perhaps, of one to three hundred thousand colonies — to store the volume currently impounded in Lake Mead. Given what we know about historic beaver abundance in North America, it's hardly far-fetched to imagine that many rodents prospering in the vast Colorado Basin. If — or, more likely, when — declining volume or sedimentation renders titanic dams obsolete, it's comforting to think that beavers could at least partly make up the shortfall.

What's more, the West isn't just debating how to cope with its aging dams — it's also contemplating new ones. Haunted by the specter of climate-fueled drought, public agencies are considering massive new water storage projects in California, Utah, Colorado, and other western states. And that's where beavers could really prove their worth.

Take, for instance, the Bear River, which rises in the Uintas Mountains, wends through Wyoming and Idaho, and finally returns to Utah to empty into the Great Salt Lake. As I write, the state's Division of Water Resources is contemplating new reservoirs in the Bear's basin; two prospective sites, which lie in the Little Bear–Logan River sub-watershed, would together store forty thousand acre-feet and cost up to five hundred million dollars.

In 2016 Konrad Hafen, a graduate student at Utah State University, decided to find out whether beavers could beat giant concrete walls. He calculated that the Little Bear–Logan watershed could support around thirty-seven hundred beaver dams, which could capture up to seventy-five hundred acre-feet of water — an impressive figure, but less than 20 percent of the proposed human-built reservoirs.[45] So beavers lose, right?

Well, not so fast. Hafen, befitting a diligent scientist, used some conservative assumptions: namely, that each beaver dam would store only a single acre-foot of surface water, and that they'd hold just as much water aboveground as below. Fortunately, we cavalier journalists can play a bit looser with our postulations. The Lands Council study I cited above, for instance, suggested that beaver ponds in eastern Washington store an average of three and a half acre-feet of surface water alone, and at least *five times* that figure belowground. Apply those figures to the Little Bear–Logan watershed, and suddenly its beavers can trap around sixty-four thousand acre-feet — 50 percent more than concrete walls. Whatever numbers you use, it's clear that encouraging our rodent partners can put a dent in our water storage needs — and can do it without impeding fish passage, screwing with sediment rates, or costing the public millions of dollars.

If everything goes according to James Rogers's plans, the Winecup Gamble Ranch might someday provide a half-decent model for beaver-based storage. Rebuilding the blown-out Twenty-One Mile Dam was beyond the Winecup's means, he told me that evening, just before we settled down to some of the tastiest steaks I'd ever eaten. Besides, it didn't make much hydrological sense. Twenty-One Mile Dam's failure had been a disaster, yes, but Rogers also considered it an opportunity to rethink his ranch's relationship with water. "We're looking again at this thing and going, What if we held that water in different places?" Rogers said. "What if we broke it up and held 1,000 acre-feet over here, and 1,000 there, and 500 feet here, and 250 over there? What if we could strategically build two or three or four or five wetlands, or recharge pools, or whatever you want to call them, where we know the water always is? And what if the beaver could help us do that?"

Art Parola, a Kentucky-based engineer, is one of the people tasked with executing that radical vision. When I spoke to Parola over the phone in November 2017, he hadn't yet finalized blueprints for the Winecup Gamble, and seemed slightly daunted by the ranch's bizarre hydrology. The essentials, though, resembled the Beaver Dam Analogue project at

Bridge Creek, in concept if not specifics. By building a few small earthen dams in incised channels around the ranch, Parola hoped to create conditions in which beavers could thrive, force water onto floodplains, recharge the Winecup's groundwater supply, and turn annual streams perennial.

"We're trying to provide the minimal amount of structure we need to let the beavers go to work," Parola told me. "You look at some of our restoration sites elsewhere, and people say it doesn't look like we actually *did* anything — it just looks like a stream-and-wetland complex doing its thing. That's what we want. We'd love to extend the areas where beavers can be, to see hundreds of acres on the Winecup turn into stream-and-wetland complexes." Parola and Rogers were discussing nothing less, I realized, than beaverizing the water system of a parcel of land the size of a small northeastern state.

To be sure, the Winecup's immensity makes it more amenable to a beaver-based strategy than other ranches. Rogers directs his fifty-four hundred black Angus cows from pasture to pasture in a vast migration, preventing the herds from annihilating the willow in any single meadow. While the Winecup's immensity is a luxury, Rogers insisted that small ranches, too, can coexist with aquatic rodents. "Even my dad's kind of won over," Rogers told me. "He fenced off his riparian, and he's not blowing up beavers anymore. He still gets mad when they get into his irrigation. Two years ago, I called mom, and she said, Yeah, your dad just took the gun, and he's going out to kill a beaver. And I was like, Hey, you tell him he owes me a phone call when he gets back." Rogers laughed. "It might take awhile, but he's getting more open-minded."

Both the Winecup Gamble's story and the Susie Creek saga illustrate an important point: that beaver management is, at bottom, land management. Absent sound ranching practices, the rodents can't return. If the arid West ever hopes to address its water problems with a beaver-based solution, it will have to alleviate riparian grazing pressure. All the Methow-style relocation projects in the world won't do much good if there's not enough willow, aspen, and cottonwood to sustain returning rodents.

And while beavers' primary competitor is the domestic cow, *Bos taurus* doesn't own the only rumen on the range. In many places, the rodents find themselves pitted against an overabundance of wild grazers — and the answer to that problem may be one that much of society doesn't want to hear.

—CHAPTER EIGHT—

Wolftopia

One frigid day in January 1995, five federal employees lugged a hulking crate, its steel sides perforated with airholes, through shin-deep Yellowstone snowdrifts above the Lamar River. Photographers milled about, cameras whirring, breath white in the gelid air. The crate's handlers were drawn from the highest ranks of officialdom: Park superintendent Michael Finley manned one corner; US Fish and Wildlife Service director Mollie Beattie took another; and Bruce Babbitt, President Clinton's interior secretary, supported the left flank. Though they resembled pallbearers, their cargo was very much alive. Inside the box lay a ninety-nine-pound gray wolf, recently captured in Canada—the first member of her species released into Yellowstone after a seventy-year absence from the park. Babbitt pressed his eye to an airhole. "The gorgeous, large gray female was lying there, alert, very calm, sort of like my dog at the dinner table," the secretary recounted later.[1]

Over the next year, the government released sixty-six wolves into Yellowstone and central Idaho. The carnivores, surrounded by prey and unconstrained by rival packs, flourished and dispersed. Within two decades more than seventeen hundred wolves roamed Montana, Wyoming, and Idaho, and intrepid bands had penetrated Oregon and Washington.[2] Along the way the predators incurred the wrath of ranchers and elk hunters, becoming the most controversial wild animals in the United States. To this day you can't travel the rural Mountain West without spotting the famous bumper sticker: a wolf in the crosshairs of a rifle sight, alongside the slogan SMOKE A PACK A DAY.

Outside Yellowstone National Park, wolves were often treated as pariahs. Within its borders, however, they were considered saviors.

If you care one whit about wildlife, you've probably seen the YouTube hagiography "How Wolves Change Rivers." If you're not among its thirty-seven million viewers, here's the gist: Before the return of Yellowstone's wolves, the story goes, unchecked herds of elk grazed the park's riparian plants to nubbins. Denuded riverbanks slumped into their channels, leaving behind bare, incised, eroding waterways. Just as unmanaged cows once ruined Susie Creek, fearless elk ran roughshod over America's most iconic landscape. Overgrazing is overgrazing, whether the perpetrator is a wild ungulate or a domestic one.

The wolf resurrection changed all that. Not only did *Canis lupus* thin the herds, wolves also frightened their prey away from narrow valleys, death-traps whose tight confines made elk easy pickings — a dynamic dubbed "the ecology of fear." Safe from hungry elk, streamside aspen and willow thrived. Wildlife from flycatchers to grizzly bears returned to shelter and feed; eroding streambanks stabilized; degraded creeks transformed into deep, meandering watercourses. Wolves had apparently catalyzed a *trophic cascade*, a dynamic in which the influence of top predators — lions in Africa, dingoes in Australia, even sea stars in tide pools — ripples through food-webs, changing, in some cases, the vegetation itself. "So the wolves, small in number, transform not just the ecosystem of the Yellowstone National Park . . . but also its physical geography," enthused the video's narrator.[3]

"How Wolves Change Rivers" transfixed me when I first saw it. I wasn't the only one: I've since heard the Yellowstone wolf tale repeated at conferences, seminars, and even on the lips of baristas in Scottish fishing villages. "This story — that wolves fixed a broken Yellowstone by killing and frightening elk — is one of ecology's most famous," wrote the biologist Arthur Middleton in *The New York Times*. And it's a great story: imbued with hope, easily grasped, bespeaking the possibility that our gravest mistakes can be remedied through enlightened stewardship. We live in a world of wounds, quoth Aldo Leopold, but we can also play doctor.

There's only small problem with the vaunted wolf narrative, Middleton added: "It's not true."[4]

"Untrue" is, to my mind, too harsh. "Incomplete" might capture it better. Wolves have undoubtedly changed Yellowstone's ecosystems, and in some river valleys they've done extraordinary good. But there are other valleys the canids haven't managed to save, places that remain as degraded as they were on that January day in 1995. Yellowstone's wolves are landscape benefactors, but they're not panaceas.

So what makes the salvational story incomplete? Well, for one thing, it elides the role of another species — an equally influential animal that, like the wolf, was for decades almost entirely absent from the park. More than twenty years after wolf reintroduction, most of Yellowstone's streams are still missing their true architects.

———

More than a century ago, the Greater Yellowstone Ecosystem ranked among the finest capitals in all of Beaverland. Although mountain men pillaged the region, Yellowstone's beavers survived the fur trade unvanquished. In 1863 Walter DeLacy complained about the "numerous beaver dams" that frustrated his travels through the Madison drainage.[5] The Earl of Dunraven, exploring the Upper Yellowstone River in 1874, remarked that "all the streams are full of beaver."[6]

There was a good reason why Yellowstone was so beaver-rich: The same otherworldly geysers and hot springs that entranced the park's human visitors made it uniquely well suited to harboring aquatic mammals in winter. "These outlets, relatively clear of ice, afford unusual advantages for burrow habitations in their banks," wrote superintendent Philetus Norris in 1881, "or for the construction, in their sloughs, of the . . . brush-and-turf houses of these animals." Were it not for illegal trappers surreptitiously controlling their numbers, Norris speculated that beavers would overrun the park altogether. "Unmolested by man, who is ever their greatest enemy, the conditions here mentioned are so favorable to their safety that soon they would construct dams upon so many of the cold-water streams as literally to flood the narrow valleys, terraced slopes, and passes, and thus render the park uninhabitable for men as well as for many of the animals now within its confines."[7]

As stronger enforcement stymied poaching, the dam builders grew ever more abundant. In the 1920s Edward Warren, a biologist from Syracuse University, spent two summers surveying Yellowstone's beavers, a prodigious undertaking that entailed measuring and photographing hundreds of dams, lodges, ponds, canals, and cuttings: "everything bearing on the life history of the beaver that came to my notice."[8] Warren concentrated on the Northern Range, a dry, undulating realm of grassy valleys, sagebrush slopes, and ravines through which coursed small streams on their way to the Lamar and Yellowstone Rivers. Every defile, it seemed, was jammed with dams and ponds. One beaver colony near Camp Roosevelt often went about its business in broad daylight, "paying little or no attention to the interested observers lined up along the road above the ponds."[9] In 1927 Milton Skinner, a park naturalist, estimated Yellowstone's beaver population at ten thousand, but added that the guess was likely "very conservative."[10]

Yet the park's beaver boom was short-lived. By the time Robert Jonas, a graduate student at the University of Idaho, replicated Warren's reconnaissance in the 1950s, he found little but ruins, like an archaeologist stumbling upon the overgrown rubble of an abandoned kingdom: empty lodges, derelict dams, chew scars darkened by the passage of years. "Of all the areas thoroughly investigated by Warren and found to have significant beaver workings," Jonas wrote, "none had any activity in 1953 nor was there any indication that beaver had inhabited those regions for several years."[11] Just three decades earlier Yellowstone had been castor central. Now, despite all the protections afforded by a national park, it was a ghost town. What had gone wrong?

Beavers, Jonas realized, were collateral damage — the victims of backward wildlife management. In 1914 Congress had granted the Bureau of Biological Survey, an obscure agency tasked with controlling crop-eating birds and rodents, a sinister new remit: the destruction of wolves, coyotes, cougars, and other predators. Sheep and cattle were spreading across the West, and the government, at the behest of anxious ranchers, sought to purge the range of stock-menacing carnivores. Bureau agents bearing guns, poisons, and traps scuttled across the land under the orders of director Vernon Bailey, a man not known for mercy. "By watching near [wolf] dens in the early morning or at dusk before the young are taken out," Bailey advised in one circular, "a good hunter is sometimes able to shoot one or both of the parents."[12]

Today national parks are strongholds for large carnivores. But when Bailey visited Yellowstone in 1915, he "found wolves common, feeding on young elk," and urged the bureau to kill "without abatement until these pests are greatly reduced in numbers."[13] The new National Park Service became one of Bailey's most enthusiastic clients. By 1926 the Park Service and the Bureau of Biological Survey had killed at least 122 Yellowstone wolves, thirteen hundred coyotes, and untold cougars.[14] The campaign was largely motivated by the cynical politics of self-preservation. By eliminating predators, the Park Service hoped to reassure ranchers and Congress that future national parks wouldn't threaten livestock. And without pesky wolves around, Yellowstone's burgeoning elk, deer, and pronghorn herds would, in theory, spill out of the park, satisfying hunters. "We have had the support of game associations only on the basis that the park would act as reservoirs for the game," superintendent Roger Toll opined in 1932. "To me a herd of antelope and deer is more valuable than a herd of coyotes."[15]

Public pressure eventually ended the slaughter, but by then it was too late: The wolves were gone. Almost immediately researchers realized how badly the policy had backfired. Exploding elk herds devoured vegetation as fast as it could grow, hastening soil erosion and forcing out deer and pronghorn. "The range was in deplorable condition when we first saw it, and its deterioration has been progressing steadily since then," a team of visiting scientists cautioned in 1934.[16]

At first, the anti-predator campaign may have given beavers a boost. Wolves are inordinately fond of the delectable rodents: When scientists picked through summertime wolf scat in Alberta, they found nearly 60 percent of the samples contained traces of luckless beaver.[17] (While hardly anyone has observed an attack, there's every reason to believe that clashes between the toothy combatants are epic: One graphic study of wolf-beaver encounters in Minnesota described "evidence of a struggle as a downed log had been torn apart on one end with claw and/or tooth marks present in the wood."[18]) In 1926 Edward Warren suggested that carnivore extermination had unleashed "what is probably an unnatural expansion of the beaver population."[19]

Soon after the final wolf fell, however, beavers found themselves squeezed out by other herbivores. Beavers in northern latitudes, recall, cache food in frozen ponds for the winter. According to ecologist Bruce Baker, the rodents are prudent about their stores, ignoring willow stems until the plants are large enough to furnish an adequate winter supply — around three years of growth. Beaver-chewed willows tend to *coppice*, growing back shoots after each cutting; beavers often harvest and rotate their coppices as diligently as any silviculturist. "There's a rest period built into the system," Baker told me.

Elk, by contrast, prefer aspen and willow at their youngest, greenest, and most tender. Nibbled incessantly by elk, willow can't recover from browsing; eventually the plants die, depriving beavers of their winter larder. What's more, the defoliated, trampled range likely exacerbated erosion and runoff, filling in beaver ponds with sediment at breakneck speeds. Wet meadows dried out; ponds became pastures. In his 1955 report Bob Jonas didn't pull punches: "The serious reduction of the favorable beaver habitat within the park boundaries can be attributed primarily to the overpopulation of elk."[20] Beaverland had ceded to Elktown.

The Park Service grappled fitfully with the antlered ravagers, angering both animal-lovers and hunters by culling more than thirteen thousand elk from 1949 to 1968.[21] But the range languished, and beaver populations remained suppressed — much to the detriment of the park's many wetland-dependent creatures. Amphibians like Columbia spotted frogs and boreal toads must have suffered, along with waterfowl such as pintails, snow geese, and trumpeter swans. The worst casualties may have been the park's dozens of passerine birds, including flycatchers, warblers, vireos, and waxwings, whose perching and nesting trees were mowed down by elk as soon as they pushed shoots above soil.

In the mid-1980s beavers finally found a champion. Dan Tyers, today a grizzly bear biologist for the US Forest Service, grew up around Yellowstone, the son of a Park Service ranger. Among his father's colleagues was Robert Jonas, the scientist who demonstrated that Yellowstone's elk were outcompeting its beavers. Jonas became a mentor to the younger Tyers, bending his ear with theories about ungulates and rodents. When, in 1978, Tyers landed a backcountry ranger gig in the breathtaking Absaroka-Beartooth Wilderness, the block of land just north of Yellowstone, he kept his eyes peeled for beavers. He saw ancient lodges and overgrown dams everywhere, but scant recent sign.

As Tyers ascended the Forest Service's ranks, the shine wore off his ranger job. He grew disenchanted with the mundanity of checking permits and monitoring timber sales. He yearned to feel creative, to ask grand questions and seek meaningful answers. His supervisors were perplexed by his zeal, but they didn't object when he proposed his big idea: reintroducing beavers into the Absaroka-Beartooth Wilderness.

Tyers's effort, which lasted from 1986 to 1999, remains one of the largest beaver relocations ever undertaken. Long before the Methow Beaver Project, he anticipated its structure, transplanting nuisance beavers from private property to public headwaters. (Tyers even trapped on Ted Turner's ranch after beavers lopped down the magnate's shade trees.) He moved the animals on horseback, in canvas-wrapped cages cooled by blocks of ice. Tyers turned his beavers loose in streams like Slough Creek and Buffalo Creek, waterways that flow south from the Absaroka-Beartooth into Yellowstone National Park, passing from Forest Service land to National Park Service jurisdiction as they go. Tyers knew there was a chance — indeed, a likelihood — that his beavers would follow the creeks into the park. Although he didn't tell me as much, I sensed that returning beavers to the places Bob Jonas studied was an ulterior motive, a way of righting the ecological wrongs that his mentor once documented.

Sure enough, Tyers's beavers soon trespassed. In 1996 Doug Smith, Yellowstone's wolf biologist, surveyed its beaver colonies from the sky, his lanky frame folded up in the back of a circling Piper Super Cub, stopping just before hitting "observer fatigue" — a euphemism, he wrote later, for "the puking point."[22] During that first survey Smith counted only forty-nine colonies in the entirety of Yellowstone. By 2007, however, the tally had nearly tripled, to 127 colonies. The spike, Smith noticed, was concentrated in the Northern Range, on many of the same streams where the Forest Service had released beavers. At last, Tyers figured, his wards had drifted downstream.

In the popular imagination, however, Tyers's relocation tale lost out to a more compelling narrative: that, by allowing willows to regrow, wolves alone brought beavers back. Articles in *National Geographic*,[23] *The New York Times*,[24] and *Orion Magazine*[25] attributed beaver recovery to wolf reintroduction, without once mentioning the relocation program that had quietly restocked the rodents just outside the park. Beavers became another link in the trophic cascade. This framing wasn't inaccurate, per se, but it omitted Tyers's meddling.

"The conversation made me think about spontaneous generation," Tyers told me when I visited his Bozeman office, referring to the antiquarian theory that, say, maggots could simply arise, without progenitors, from rotting meat. "How did beaver get there? Well, they just appeared.'" He sat back in his rolling chair and shrugged. "Beavers started showing up in Yellowstone, and because of the park's name recognition it hit the news. Reasonably so — I'm not being critical. But all the while we'd been moving beavers into the backcountry and probably had fifty different active lodges, and terraced dams that went on for a mile. Holy smokes, folks, if you just went a few miles north of the park and saw what beavers were doing *there*, you'd have a different view of the world."

The media coverage glossed over another inconvenient truth: Despite the wolf reintroduction, and despite beavers' incremental recovery, many Yellowstone valleys remained untransformed. Had Edward Warren surveyed the Northern Range in 2006, he would have discovered that the soggy paradise he'd chronicled in the 1920s remained largely bone-dry and beaverless. The sad reality, some scientists suspected, was that a century of mismanagement had inflicted damage that neither wolves nor beavers could readily repair.

———

I have spent more time tromping through Yellowstone National Park than I have just about anywhere else. Soon after graduating college, I drove west to work for the Park Service's fish program, a grisly, slimy, endlessly entertaining position that entailed killing Yellowstone's invasive trout so that native cutthroat trout could thrive. Every week found my small team in another remote district on another memorable adventure: hiking up Specimen Creek with vats of poison strapped to our backs; sloshing through Soda Butte Creek electroshocking brook trout; floating down the Madison River at 2 AM through kaleidoscopic snowfall, fingers too numb to function, our rubber raft's headlights illuminating elk splashing through downstream crossings. I have returned to Yellowstone nearly every year since; in 2016, the week after our wedding, Elise and I honeymooned eight miles up Slough Creek. I have crossed paths with grizzly bears on Mount Washburn, smelled the potent musk of bull elk in the Thorofare, and watched a wolf squat to defecate in a parking lot near Yellowstone Lake, looking for all the world like a domestic dog.

All of this is to say that Yellowstone is one of the places I assumed I knew best — and yet, until the summer of 2017, the park's beavers were utterly invisible to me.

One June morning Dan Kotter and Lewis Messner drove me into Yellowstone National Park to rectify my oversight. We left early from Gardiner, Montana, the fetching tourist town at Yellowstone's northwest gateway, coffees riding in cupholders, Springsteen humming through car speakers. We bypassed Roosevelt Arch, the grandiose stone span that marks the entrance, and chugged uphill through Mammoth Hot Springs, where laconic elk lounged outside the cafeteria. Kotter cut west into the Northern Range, its sagebrush plateaus freckled by dark bison and umber pronghorn. At last we sidled onto a gravel pullout shaded by lodgepole. We'd reached Elk Creek, Exhibit A in the case for the importance of Yellowstone's beavers.

Kotter, a PhD student at Colorado State University, led me into the valley, while Messner, a master's student in the same department, paused at a PVC pipe, thrusting from the soil like a plastic sapling, to record groundwater levels. At first glance there wasn't much wrong with Elk Creek, which tinkled gaily through a meadow of smooth brome and timothy grass. I'd driven past it a hundred times without casting a second glance. Not until we climbed down into the streambed did I recognize how severely Elk Creek had incised. A towering cross section of eroding bank loomed eight feet above the ankle-deep channel, black and rich as chocolate cake. The stream had knifed through its floodplain to bedrock.

The scene was even more depressing when you realized what preceded it. When Edward Warren surveyed Elk Creek in the 1920s, he found a wonderland of open water, willow, and aspen; in the stream's North Fork, he recorded seventeen dams, including "a splendid new structure . . . more than 350 feet if measured along all the curves, and 5 feet high on the lower face."[26] But the concerted wolf-killing had allowed elk to feast here, outcompeting beavers for precious forage. Absent plants and aquatic engineers, the waterway disconnected from its floodplain, devolving from productive wetland to fallow pasture. Millennia to create, decades to unravel.

Now hardly a willow stem adorned the channel: The water table had plummeted so far, Kotter said, that their roots couldn't tap groundwater. "The only thing that could re-create this big wet valley would be beavers coming back," Kotter explained. "But can beavers come back when there's no food resource?" He shrugged. "I don't think this site will restore itself in my lifetime."

This was not the Yellowstone narrative that I was accustomed to hearing. Hadn't the return of wolves allowed willow to recover? In some places, Kotter agreed, that seemed the case. Other valleys, however, like Elk Creek, had crossed a Rubicon of degradation. "I believe that wolves have done a lot of good," Kotter said as the sad thread of water wound past us, locked deep in its own basement. "But it also seems like the ecosystem in some river valleys has shifted to an alternate stable state, and it will take beavers to push it out."

Kotter is a sturdy, thoughtful guy, so soft-spoken that when we first met at a bar in Gardiner I strained to hear him over the thrum of ambient conversation. He has a mountain man's tolerance for hardship: During a winter stint in Glacier National Park, he tracked wolverines alone, on skis, in subzero temperatures; the next summer he pepper-sprayed an angry grizzly in the face as she swatted at his leg. Unlike Jim Bridger, though, Kotter is fluent in scientific jargon, sometimes to the point of comedy: "Excellent example of a zoogeomorphic agent," he remarked, straight-facedly, as we later watched water bubble from a squirrel hole. When the blackflies aren't too thick, he's fond of perching atop a rock and observing bison wallow for hours, lost in thought, dreaming up the next phase in his research. Often, he told me, tourists will see him admiring a roadside stand of Bebb willow and ask what he's ogling, assuming it's a bear or wolf. Voice full of wonder, he'll respond, "You *have* to see this *Salix bebbiana*!"

Kotter's combination of woods sense and scientific acumen make him the perfect heir to the Elk Creek experiment, one of Yellowstone's most revealing research projects. From the incised bank, we traipsed a quarter mile downstream, arriving at a stand of soaring, impenetrable willows waving their spindly arms twelve feet in the air — an impressive contrast with the browsed-to-nubbins plants I'd seen elsewhere in the Northern Range. A mesh fence, tall enough to exclude even the longest-necked elk, surrounded the thicket. Catbirds mewled in the brush. The fence, Kotter explained, simulated the effects of wolves by preventing ungulates from heedlessly devouring willows. No surprises there: Thwart elk from browsing, and plants grow bigger.

But the fence was only half the story. Kotter led me to the stream channel, where two small, human-built check dams — ungainly structures cobbled from logs, rocks, and black plastic tarps — pooled water. While the fence was designed to mimic wolves, the dams replicated beavers. Put crudely, the Elk Creek experiment was designed to answer a seemingly simple question: Which keystone mammal did Yellowstone miss more, the apex predator or the hydraulic engineer?

That problem had occupied the mind of David Cooper, a white-bearded ecologist at Colorado State, since the late 1990s, several years after wolves' triumphant return to Yellowstone. Among ecology's hoariest debates is whether ecosystems are primarily controlled from the top down, by predators, or from the bottom up, by fundamental resources like nutrients. Cooper wanted to know whether Yellowstone was regulated by its new wolves or by the most important building block of all — water. By setting up both wolf-mimicking exclosures and beaver-imitating dams, Cooper could, in theory, determine whether overgrazing or depleted groundwater was limiting willow growth. (Not surprisingly, Yellowstone's authorities frown on dam construction within the park; Cooper circumvented the prohibition by dubbing his structures "flow velocity inhibitors.") "I believed the original story — that wolves were going to chase all the animals away, all the willows were going to grow, and we were just going to monitor the response," Cooper told me later. "The way things went instead was a good learning experience for us about how complicated ecosystems are."

Beginning in 2001, a succession of graduate students measured tens of thousands of willow stems as they grew, leafed, died, and sprouted back. Holding up a set of calipers to willow after willow, ad infinitum, proved as tedious as you'd think; when I asked Dan Kotter if he'd named any of his plants, he grimaced and said, "Most of them are curse words." The results, however, were worth the trouble. In one eye-opening 2013 paper, Cooper's team found that the plants that fared best, by far, were the ones whose roots were moistened by the artificial dams. Willows that were merely surrounded by fences, but didn't get sub-irrigated by imitation beaver ponds, grew only marginally taller than ones that didn't receive any protection at all.[27] Wolves couldn't recover willows alone, in other words — they required help from beavers. "Without beavers and the high water tables they create, these riparian systems as we know them can't exist," Cooper told me.

That's unfortunate, because Yellowstone is still a challenging place to be a beaver. After our Elk Creek tour, Kotter and Messner took me to nearby Crystal Creek, one of the handful of Northern Range streams where beavers are truly flourishing. Compared with Elk Creek, Crystal looked like rodent paradise. Beavers moved here in 2015, Kotter told me, and swiftly transformed the willow clutter into an open, light-filled compound of ponds, side channels, and terraced dams. (Kotter, a classic rock fan, named the beavers at his study sites after characters from Beatles songs: Michelle, Lucy, Dr. Robert, and, at Crystal Creek, Martha. The lyrics to the song "Martha My Dear" — "When you find yourself in the thick of it / Help yourself to a bit of what is all around you" — contained, Kotter pointed out, an all-too-appropriate maxim for a beaver living in a willow thicket.) Now the channel spanned the width of a museum gallery, and a conical island of brush — the Lodge Mahal, Kotter called it — thrust proudly from its ponded heart. Swallows pinwheeled above the complex, skimming low like crop dusters to sip emergent mayflies.

For all Crystal Creek's beavery glory, however, Kotter was not bullish about the colony's long-term prospects. In just two years the Lodge Mahal's residents had drastically depleted their willow stores. In part, that's because the usual cycle of coppicing and regrowth had been short-circuited by an herbivore even more voracious than elk: bison. Between 2000 and 2016 the park's bison population more than doubled, from twenty-six hundred to fifty-four hundred — partly because the shaggy beasts have filled elk's niche, and partly owing to Montana state policies that prevent bison from leaving the park for fear that they'll spread disease to cattle. Wolves generally leave the tank-sized grazers alone, meaning there's nothing stopping bison from inhaling willow, aspen, and cottonwood shoots. At Crystal Creek thousands of their cloven hoofprints scurried over the muddy ground, divots half filled by water. Kotter hefted a dead plant that looked like it had been uprooted by a Rototiller. "Bison don't nibble daintily on a shoot like an elk," he said. "It's more like Jabba the Hutt eating a piece of pizza."

Whether willow could grow fast enough to survive depended in large measure on whether Crystal's beavers, and the elevated water tables they created, could stick around. In Kotter's and Cooper's version of the story, much of Yellowstone remains trapped in a feedback loop of lowered water tables, stunted plants, and degraded streams that not even wolves can disrupt. As Cooper and colleagues summed it up in one paper: "The absence of beaver opposes the return of tall willows and the absence of tall willows opposes the return of beaver."[28]

"Can beavers return to the Northern Range? Yes," Kotter said gravely. "Can they persist? I don't know."

———

This version of the story — that many Yellowstone valleys are too far gone for wolves to rehabilitate — has, it must be said, many smart and fervent critics. Among them is Doug Smith, the rugged, Fu Manchu–sporting wolf biologist who, every other year, crams himself into a light aircraft to survey the Northern Range's beaver colonies. Smith earned his PhD studying beavers in Minnesota's Voyageurs National Park and has spent much of his career contemplating the similarities between his two great mammalian passions. "One's an herbivore, and one's a carnivore, but beavers are just like wolves," he told me. Both species, he pointed out, are cooperative breeders, meaning that the entire family pitches in to raise their offspring. Both species jealously guard territories from rivals. The primary difference is that, while Yellowstone's wolves are probably the most famous carnivores in the world, its beavers have remained obscure. "When I'm doing wolf research, I always attract larger numbers of people," he said ruefully. "With beavers, I'm basically by myself out there."

Both wolves and beavers, of course, also have reputations as keystone species. Smith has witnessed their influence firsthand. When he arrived in Yellowstone in 1994, just before wolf reintroduction, every stem along Crystal Creek suffered from browsing. "Now you go over there, and it's a cacophony of birdsong," he told me. Willow-dependent species like Wilson's warblers are returning to other valleys. I asked Smith what had changed, and he answered without hesitation: "It's the trophic cascade."

Smith's certainty is grounded not only in personal observation but in data. Several months after my visit to Yellowstone, I spoke with Bob Beschta, an Oregon State University hydrologist who's among the leading proponents of the trophic cascade theory. Since 2003 Beschta and his frequent coauthor, the ecologist William Ripple, have published around twenty studies demonstrating the renaissance of the park's vegetation. They've measured aspen, cottonwood, alder, willow, serviceberry — if it has leaves and grows near streams, Ripple and Beschta have studied it. Most of their findings adhere to a telltale pattern: Plants have grown taller since wolf reintroduction, and they've fared better in the narrow valleys that elk, in theory, have increasingly avoided. In 2016 the two scientists examined twenty-four papers on Yellowstone's riparian vegetation in a meta-analysis — a study of studies. All but two studies showed that plant growth had increased.[29]

"At first it was sparse and local — plants were just beginning to creep up. People said it wasn't happening," Beschta told me. "Now it's clear that a broad suite of species in Yellowstone is beginning to do better, and we're just at the beginning. It's pretty amazing." Like Smith, Beschta was unequivocal in assigning credit for the transformation. "If it wasn't for wolves and other predators pulling back the ungulate pressure," he said, "I don't care *what* the hydrology was doing. Willows would not have been able to grow."

That wolves have reduced elk population densities, there can be little doubt. Whether the carnivores are actually changing elk *behavior*, though — whether the vaunted "ecology of fear" is truly in operation — is a tougher question. Researchers at Utah State University have found that adult elk are seldom targeted by wolves; even the fiercest canines aren't too keen to tangle with a huge, aggressive, sharp-hoofed beast.[30] And whether wolves alter elk populations and movements enough to generate a widespread plant resurgence, a beaver revival, and the transformation of stream channels — well, those are the hardest questions of all.

We may soon learn the depth of wolf influence on a vast scale. Since their reintroduction to Yellowstone and Idaho, wolves have trekked down from their Northern Rockies redoubts and spread west — over the Bitterroots, through the Sawtooths, across the Snake River, and into Oregon and Washington. As I type, there are wolves in the southern Cascades, wolves in Northern California, wolves within seventy-five miles of Seattle.

Among the many observers heartened by their return is Suzanne Fouty, a recently retired Forest Service hydrologist based in arid eastern Oregon. If you set up a Venn diagram whose circles were wolf admirers and Beaver Believers, Fouty would lie at its overlapping center. Her aquatic rodent preoccupation began as a PhD student at the University of Oregon in the late 1990s, when she attempted to study how fencing out grazers changed the contours of streams in Arizona and Montana. Like so many scientists before and after her, she was initially dismayed to find beavers meddling with the integrity of her research sites. A week passed before she realized that beavers were no mere confounding variable — they were the most compelling narrative thread in her streams' stories. In some cases, she found, the builders captured enough sediment to reconnect incised creeks to their surrounding valleys in under a year.[31] "When you take off grazing pressure from livestock, elk, and deer, you get a significant vegetation recovery," Fouty told me. "But you don't take the next step until you get beaver back in the system."

If uncurbed grazing is the disease that's suppressing beavers on Western rangelands, wolves, Fouty believes, represent the medicine. Persuaded by evidence from Yellowstone and her own observations, Fouty has become an outspoken proponent of the carnivores' ability to scatter wild grazers, entice beavers, and rehydrate rangelands. "The point I'm trying to get across is that we have to have wolves to have beaver, and we have to have beaver to have water," Fouty told me. "For a long time, ranchers would say, it's not the cows' fault, it's the elk. So now people like me are saying, okay, yes, the elk are a big problem — that's why we need wolves now." She sighed. "That's my pitch. And what I get is, 'We have no clue what you're talking about.'"

Although wolves have returned to the West to stay, whether they can truly flourish is a knottier question. Time and again, the canines unwittingly trigger a brutal cycle: Wolves kill livestock; ranchers complain to the government; the government kills wolves. In 2016 Washington State biologists gunned down most of the Profanity Peak pack by helicopter, while Oregon eliminated four members of the Imnaha pack, including its alpha male. The exterminations have taken their toll: After years of growth, Oregon's wolf population appears to have stagnated, and the number of breeding pairs has begun to decline.[32]

"It's frustrating because every time you take those wolves out, you're starting over again," Fouty said. "You never get the expansion you need, you never get the recovery you need, and then you go into an extended drought and your systems are completely nonfunctional. And that's not because of the climate — it's because of the decisions agencies have been making."

Oregon has an equally tragic relationship with beavers. *Castor canadensis* is the state's official animal, the mascot of its largest university, and the supplier of the Beaver State's nickname. When, in 1849, the Oregon Territory defied the federal government by creating its own mint, beavers featured on the coins; today, a golden beaver, waddling across a navy field, adorns one side of Oregon's flag. Yet for all its iconography, the Beaver State remains notably hostile to its eponymous rodents, classifying the herbivores, ludicrously, as predators. According to records Fouty collected, between 2000 and 2015 the state's Department of Fish and Wildlife and federal Wildlife Services reported 53,983 beavers slain in Oregon — around 10 rodents per day. Often, the killings seemed to defy common sense: From 2004 to 2007, for instance, as eastern Oregon grappled with disastrous drought, 262 beavers were eliminated in bone-dry Baker County alone.[33]

"And it's not just the beavers killed," Fouty added. "It's all the progeny they never had. I think it's inexcusable that we're not doing anything to protect our communities from drought." The future of water on America's rangelands may rest on the outsized influence of two keystone species: one rodent, one canine, neither welcome.

———

The day after our Elk Creek trip, Dan Kotter, Lewis Messner, and I hiked eight miles up Slough Creek, a stream that horseshoes lazily through sweeping bison meadows on its way into the park from the Absaroka-Beartooth Wilderness. Twenty thousand years ago the Pinedale Glaciation had gouged out this valley with a flick of its icy finger. Over millennia, Kotter said as we climbed, beavers aggraded the ice-carved trench into the lush grassland below us, filling a vertiginous V into a hospitable U. We could still make out the grass-covered shadows of ancient dams snaking across the damp meadows.

Now Slough Creek's beaver population was again growing. When Doug Smith first surveyed the park's beavers from the air in 1996, Slough Creek hadn't supported a single colony; a decade later it held nine. We met beaver sign everywhere: foraging canals slicing through damp floodplains, gnaw marks inscribed into small spruces, spindrifts of polished sticks washed ashore on point bars. I suggested Kotter name a Slough Creek beaver after Eleanor Rigby; he promised to take it under consideration. "This place," he said, "is the beating heart of Northern Range beaver recovery."

At last we stumbled upon shelter, a charmingly dilapidated cabin perched at Slough Creek's confluence with a rollicking tributary called the Elk Tongue. No sooner had we dropped our packs than we were passed by a beaver coasting down Slough at high velocity, mocha head poking from the current. Later, we saw the same animal struggling back upstream at a much slower clip, riding high and ungainly, like a paddling golden retriever. It was the first time I'd seen a beaver look clumsy in his aquatic element. Kotter pointed out, though, that the creek's flow was weakest at the surface, where it was diminished by friction with the air. Even awkward-seeming beaver behavior, I realized, conceals efficiency.

After dinner we set out for some evening recon. Flooding had left Slough's meadows soft and saturated; soon we were soaked to the shins. Baseball-sized boreal toads, a species that in many places breeds exclusively in beaver ponds, leapt to avoid our footfalls. The sun vanished behind a valley wall, casting a blush of alpenglow upon the snow-dusted Beartooths. Sandhill cranes warbled eerie good nights; a twisting trio of ducks transcribed a corkscrew overhead.

Although Slough Creek's dams had been obliterated, their creators hadn't gone anywhere: Every quarter mile a haystack-sized lodge sprouted from the bank. Kotter cut across the floodplain to examine a massive dome with a freshly manicured roof. We knelt to examine tracks, a jumble of splayed hind feet and dainty front paws. Some of the impressions were no wider than a thumbprint. "I think there are kits in here," Kotter whispered. We held our breath, and moments later faint burbles, uncannily like the cries of a human baby, rose from the lodge's interior.

Evidently the parents did not appreciate our proximity, for the chirrups were followed by a resounding *kerplunk*, as though someone had thrown a flagstone into the stream. A glistening black head cut a wake through Slough Creek, whose glowing surface reflected the fiery sunset. The beaver swam toward us with startling boldness, tacking back and forth upstream like a sailboat battling a strong wind. He whacked his oar-like tail again, shattering the dusk and raising a plume of spray. He was brave, determined, selfless in the face of danger; I felt instantly guilty at causing him undue stress.

"There's our Sergeant Pepper," Kotter whispered.

As we retreated through near-darkness to the cabin, bear spray at the ready, I realized this encounter represented something new to me. In my beaver-watching career, I had seen the adaptable animals flourishing in irrigation ditches, culverts, and drainage canals. But I'd never met one in a place so wild that I could gaze from valley wall to valley wall, upstream and down, without laying eyes on house, road, or artificial light. We have countless examples of how *Castor canadensis* uses and abuses human-built infrastructure, but precious few places where we can observe beavers interact with a full complement of native wildlife — wolves and elk, bison and boreal toads, cutthroat and cranes. Beavers are defined by their role as keystone species; the Greater Yellowstone is one of a handful of ecosystems where the arch remains intact.

Even so, there's something fallacious about calling Yellowstone wild. Over the last 150 years, its denizens have been hardly more self-willed than the orcas at SeaWorld. Wolves were wiped out by one government agency, then reintroduced by another; elk were culled and subjected to firing squads the moment they crossed park boundaries. Bison still face hazing or slaughter when they summon the temerity to leave Yellowstone. Even the return of beavers, one of the park's happiest stories, was abetted by a relocation program engineered by an idealistic Forest Service employee. Our wildest ecosystems are indelibly smudged with human fingerprints.

As long as beavers are back in Yellowstone, why does it matter who, or what, gets credit for their return? Ask Dan Kotter, and he'll tell you the reason is expectation management. Wolves may well be boosting vegetation in many Northern Range valleys — indeed, having pored over Bob Beschta's data, I have little doubt they are. Yet some incised streams, like Elk Creek, have degraded too far to bounce back, locked today in the dreaded purgatory that Kotter calls the "alternate stable state." For all the good that wolves do, asking them to be cure-alls might be a demand too far. I've thought a lot about how to reconcile the competing Yellowstone stories, and here's the best I can do: In many heavily grazed western land-scapes, restoring beavers demands wolves. But some ecosystems are too damaged for even predators to salvage.

Nor are wolves the only way to alleviate the symptoms of elk overabundance. Four hundred miles southeast of Yellowstone, in Colorado's Rocky Mountain National Park, a parallel ungulate explosion drove the park's beavers nearly to extinction. By 1980 nary a beaver cavorted in Beaver Meadows.[34] While Yellowstone tackled its elk glut with wolves, Rocky Mountain — which is practically next door to Denver — took a more cautious approach, encircling over two hundred riparian acres with six-foot-tall fences and planting thousands of aspen and willow stems. Although the park's elk population has descended to sustainable levels, beavers have recolonized only 10 percent of its suitable habitat,[35] and landscape ecologist Hanem Abouelezz told me vegetation throughout much of Rocky Mountain still isn't ready to support the rodents. "If they just mow down all our plantings and den in the bank, that could really set us back," she said. Park officials have discussed the possibility of installing artificial dams to boost water tables and willow growth, but Abouelezz told me the idea was still embryonic. "We didn't create this problem in ten years," she said, "and we're not going to fix it in ten years."

Whether it's appropriate to build artificial beaver dams in national parks is an ethical question as much as a scientific one. What do you value more: rapid recovery, or the relative naturalness of a hands-off approach? Without intervention, Yellowstone aficionados may have to wait a long time — centuries, perhaps — to glimpse the gloriously ponded Northern Range that Edward Warren sloshed through in the 1920s. But maybe that's okay. "Twenty years ago, Yellowstone was a dismal desert," Bob Beschta told me. "Today, most aspen stands are on the trajectory of success. Personally, I'm not in a huge rush to push the system. I think it's working." Installing artificial dams to accelerate natural recovery, Beschta added, would be "an ecologically bankrupt idea."

Beschta's aversion to heavy-handed meddling is also why he wishes that Dan Tyers had never relocated beavers into the Northern Range. Wolves would have paved the way for the rodents' eventual return, he insists. "He could've dumped a thousand beaver outside the park, and you'd still never have any in the Northern Range if the plants hadn't recovered," Beschta said. "The reality is there was no place for them before wolf reintroduction — zero." To Beschta's mind, the Yellowstone story proves that bringing back beavers throughout the West means keeping ungulates, both domestic and wild, in check — whether with more assertive cowboys or more abundant carnivores. "Until we resolve grazing issues, until we get functioning riparian plant communities that allow beaver to come in, we're going in the wrong direction."

———

We woke the next morning from a sound night's sleep in our Slough Creek cabin to find that rain had fallen and bejeweled the grass. After breakfast

Kotter cracked open a cabinet and unearthed a trove of old logbooks, scrawled with entries authored by visiting rangers and researchers. I sat on the wooden porch in the sun and leafed through forty years of semi-legible jottings. Some were quotidian: "The outhouse door slams at the same moment every morning." Others, dramatic: "Grizz print on door is a front pad and is 6 inches wide." I recognized the handwriting of my former fish crew supervisor, and the names of researchers I knew by reputation. There, on November 30, 1988, was Dan Tyers, up Slough Creek to radio-collar moose: "Joys of early winter skiing — breaking trail; not as bad as trip last January."

Wait a second — collaring moose? I'd been into Slough Creek half a dozen times over the years without once seeing hide or hair of *Alces alces*. As I paged through the logbooks, though, I discovered that both moose and elk had once been ubiquitous in Slough Creek. Nearly every journal entry mentioned encountering one or both species. In 1988: "Almost 200 elk seen between here and the Silver Tip ranch." In 1989: "We've seen at least eight moose, many ducks, elk with new calves." In 1990: "Saw elk in every meadow."

While browsers apparently had the run of the place, beavers, I noticed, had been scarce. An entry from June 7, 1989, signed by one of Tyers's assistants, told the tale: "Not good beaver habitat until you get to Frenchy's meadows" — a bend far upstream. Now the species' roles were reversed: Elk and moose had nearly vanished, and beavers had taken their place. The old logbooks might not qualify for publication in a peer-reviewed journal, but they still provided compelling testimony on behalf of a restructured ecosystem.

The most powerful stories tend to be the simplest ones, the tales that can be cogently distilled into four-minute YouTube clips. Ecological truth, however, is harder to condense in a viral video. Much though we crave a unified field theory for Yellowstone National Park, we may instead be forced to settle for *dozens* of stories, each unique to its stream, each the product of a different permutation of topography, hydrology, and ecology. Yellowstone is larger than some European countries; it contains multitudes. There is room for waterways, like Slough Creek, that have been transformed by predators, and for others, like Elk Creek, that no number of wolfpacks could ever restore.

"The two most common words in ecology," Dan Kotter told me as we flipped through the musty logbooks, "are *It depends*."

Toward the back of one journal, I came to an entry penned in 1991, signed by a mysterious *M. B.* — perhaps Mollie Beattie, former head of the Fish and Wildlife Service, who died of cancer in 1996, just a year after wolves were reintroduced. The log prattled on for a few sentences about meteorological conditions before wrapping up with a premonition that, in its wistful mix of hope and yearning, hit me with the force of a line from *A Sand County Almanac*. "All this place needs is a pack of wolves to serenade us to sleep," M. B. had written, four years before the carnivores' return. "We're working on it!"

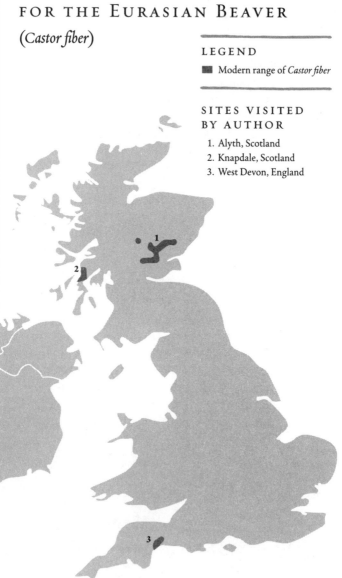

UK Reintroduction Sites for the Eurasian Beaver
(*Castor fiber*)

LEGEND

■ Modern range of *Castor fiber*

SITES VISITED BY AUTHOR

1. Alyth, Scotland
2. Knapdale, Scotland
3. West Devon, England

— CHAPTER NINE —

Across the Pond

I n the year 1232 — before Columbus landed in the Bahamas to torment the Arawak, before Marco Polo arrived in the court of Kublai Khan, fewer than twenty years after John signed the Magna Carta — King Alexander II gifted Neish Ramsay a sizable parcel of fields and woodlands in a verdant corner of eastern Scotland. Neish did much to earn his bequest: Local legend purported that he'd studied medicine under a wizard and, in the course of preparing some magical pharmaceuticals, had ingested a dollop of white snake venom, which, naturally, imbued him with X-ray vision. When the king fell ill, the physician hastened to the palace, where he diagnosed Alexander II with a hairball lodged in his innards and — via surgery, not sorcery — extracted the offending object. The grateful king gave Neish a twenty-six-hundred-acre estate in thanks, promising that the property would belong to Neish's "bairns and his bairns' bairns until the end of time." If the king had known that the estate would someday become Ground Zero for the re-beavering of the British Isles, and that one of Neish's bairns would be arrested under suspicion of unauthorized rodent release, he might have ruled differently.

Nearly eight centuries after his ancestor removed a royal hairball, Paul Ramsay, the bairn in question, led me, Elise, and a dozen curious visitors down a Perthshire country lane for an evening of beaver-watching. Chill rain had spattered the estate's grounds all afternoon, but now golden evening light poured through fleeing clouds, and the world smelled new. Bedraggled ewes watched us pass with their ears pricked, lambs pressed to their flanks.

Ramsay doglegged into a copse and we followed, weaving through rhododendrons still glittering with raindrops. Our leader picked his feet up and put them down in a cartoonish imitation of stealth, like Elmer Fudd on a wabbit-hunt. He turned to his guests and placed a finger against his lips. "We must proceed with utter quiet," Ramsay cautioned, then added, a bit cryptically, "Quiet but bold!"

After a few minutes of walking, neither very quietly nor very boldly, in single file, our party arrived at the fringe of a magical pocket wetland bordered by pines. Yellow gorse grew riotous on the hillside, mirrored in the lambent pond. Ramsay unfolded the tripod-mounted spotting scope that he carried slung over his shoulder. A beefy tourist from Wisconsin nudged his shoulder. "How often do you see them?"

Ramsay considered. "There's a lot of meditation involved," he murmured. He was seventy-one years old, with a full white beard and a gray mane that flowed from beneath a black beret that he wore at a rakish pitch. Before I met Ramsay, I generally dismissed the notion that eyes can twinkle as both cliché and anatomical impossibility. And yet his seemed to spark blue with suppressed mirth, as though the world was full of obscure jokes that only he — and his partner in crime, Louise — quite understood. He struck me as a worthy heir to another playful Scottish conservationist, John Muir.

"Eventually, something may appear," he told the cheesehead with a vague shrug. "But then again, maybe not."

We waited. Songbirds, their accented calls beyond my powers of identification, twittered in the pines, and waterfowl skated across the pond. Though the hour had long passed 9 PM, the hills remained lit with the day's embers. Elise and I scanned the surface with our binocs. Finally, they arrived from nowhere, as if summoned: two plump beavers, fur nearly black with wet.

The bystanders took turns watching through Ramsay's scope as the beavers cruised their pond, ears and nose poking above the water like periscopes, assessing the maintenance that would require their attention this coming night. Then they hauled themselves onto land for an evening snack, munching at the grass that bordered the pond, as content to graze as any sheep. I recalled Enos Mills's lovely description of beaverhood: "His lot is cast in poetic places."[1] Ramsay smiled beatifically, bearded chin in hand.

In many North American towns, this scene is so common it's mundane. Beaver-watching for us Yanks is hardly more complicated than heading down to our local pond, marsh, or river and waiting for the animals to emerge. Observing British beavers in their element, by contrast, likely became impossible sometime during the reign of George III. Although no one's positive when beavers vanished from the isles, we know that hunters had extinguished them in Scotland by the seventeenth century and in England by the late eighteenth.[2] Even as American populations rebounded from the fur trade in the early 1900s, Britain remained beaverless.

Today, thanks to conservationists' efforts, beavers have regained a clawhold in the United Kingdom. In some watersheds they've established outposts with the grudging assistance of governmental reintroduction programs. Elsewhere they've formed breeding populations with the aid of private landowners, some of whose bestiaries have elicited the ire of farmers and officials. And none of those rebel rewilders have drawn more acrimony and admiration than Paul and Louise Ramsay, laird and lady of beaver manor.

After the beavers slipped off into the gloaming, visitors, including some local Scots who'd come by to glimpse their aquatic neighbors, peppered Ramsay with questions. What did beavers eat; where had they come from; how many were wandering Bamff? Inevitably, or perhaps because Paul steered it there, the conversation turned to sex. Ramsay described the beaver's curious cloaca, related Aesop's myth of the self-castrating beavers, and imparted a brief lecture on their monogamous social structure. "I'm sorry to say that the male beaver looks to be a considerably more faithful husband than we humans," he concluded, to laughter.

Suddenly Ramsay turned serious, or mock-serious. The light was failing, yet there were more beavers to see. "We'll just try to get downwind of them," he said, resettling his jaunty beret. "We must resume our silence, like Trappist monks!"

———

Long before Europeans gobbled up the New World's furs, they annihilated their own beavers in the Old. *Castor fiber*, sister species to our *canadensis*, once inhabited nearly all of Europe, as well as much of Asia. Beavers teemed in the Rhine River, the Seine, the Elbe, the Danube. When Genghis Khan's Golden Horde floundered in the Pripyat Marshes in the thirteenth century, the impenetrable swamps were chock-full of beavers.[3] Vikings sometimes interred their dead with their effects enclosed in a beaver-skin bag.[4]

Beavers first arrived in Britain well over a million years ago, when an ice age dropped global sea levels and turned the English Channel into a land bridge. Before too long, they had bipedal company: Beaver bones appear in five-hundred-thousand-year-old middens buried by *Homo heidelbergensis* in modern-day Sussex.[5] Although glaciers repeatedly drove both hominids and castorids back to mainland Europe, beavers returned for good in 9500 BC, spreading through modern-day England, Scotland, and Wales.[6] The engineers created streamside clearings in which humans could hunt and fish, irrigated agricultural fields, and perhaps taught people how to coppice trees to generate firewood and building materials. The crests of their dams provided sturdy throughways across the isles' infamous bogs and marshes. Their half-submerged figures may even have birthed Britain's most notorious cryptid. "That most famous Scottish monster from Loch Ness . . . happens to be in that part of Scotland where there is a late record of beaver survival, and an oral tradition of their presence," wrote the archaeologist Bryony Coles in *Beavers in Britain's Past*. "Were the monsters no more than a family of beavers seen by twilight?"[7]

By the nineteenth century, however, you were about as likely to spot a beaver in Britain as you were Nessie herself. Pelt, meat, and castoreum hunters drove beavers from the country, leaving behind naught but bones, buried sticks, and an atlas of obsolete place-names: Beverley, Beversbrook, Buernes.[8] The trajectory was more gradual in mainland Europe, but hunters eventually ferreted out most of the continent's final strongholds. The demand for *Castor fiber* was spurred, in part, by the Catholic Church, which classified beavers, whales, otters, and other water-dwelling mammals as fish — making the rodents one of the few forms of red meat that parishioners could guiltlessly consume during Lent.[9] By 1900, only a thousand or so beavers hung on in Europe and Asia, confined to dwindling relict populations in France, Germany, Norway, Russia, Belarus, Ukraine, China, and Mongolia.[10] The crisis was even grimmer than North America's parallel collapse.

In the early twentieth century, the tide began to turn. Surviving colonies received official protection, and a series of haphazard relocations scattered beavers to the winds. Norwegian beavers were transplanted in Sweden, Belarusians shipped off to Estonia, French beavers plopped in Switzerland. The reintroductions took none of the Methow Beaver Project's painstaking care — one reviewer sighed that the releases generally entailed "remarkably little planning or monitoring" — and sometimes gave no thought even to species.[11] When Eurasian beavers were unavailable, biologists simply imported North American ones, unaware at the time that the continents hosted different castorids. Several populations of *Castor canadensis* now dwell in Europe, including a cluster of more than ten thousand non-natives in Finland. Luckily we don't have to worry about hybrids: The two species have different numbers of chromosomes, and all interbreeding attempts have failed.[12]

But it wasn't until the 1990s that European beaver reintroduction kicked into overdrive, pushed by *der alte Bibermeister*, the German king of castors: a man named Gerhard Schwab.

European beaver-lovers comprise a small and tight-knit tribe, and, by fortunate coincidence, Schwab happened to be visiting the Bamff Estate at the same time as Elise and me. One morning he permitted me into his sunlit guest quarters. Schwab, a mountainous man with a beard and ponytail as thick and snowy as polar bear fur, began our conversation by gifting me two furry beaver keychains. He waved off my thanks. "We had three thousand of these made," he said. "I give them away everywhere I go."

Schwab's story begins in 1988 when he returned to Germany after receiving a master's in wildlife biology at Colorado State University. Since western Europe lacked the large mammals that he longed to study, he fell into a job managing beavers in Bavaria, the southeast German state that includes Munich. As Bavaria's *Biber* numbers swelled in the '80s and '90s, the rodents dispersed from big rivers into drainage ditches alongside arable land. At night they stole into fields and munched on corn and sugar beets.[13] Although Schwab's research suggested that most Bavarian beavers stayed out of trouble — "who cares if they're eating a few marks of corn?" — he concluded that some mischief-makers should be trapped. German environmental groups revolted. "They've been extinct for almost a century," Schwab laughed as he recalled the consternation of wildlife advocates, "and you bring them back and start killing them again!"

The impasse was cleared by a colleague who'd studied the possibility of relocating beavers to Croatia. Could Schwab trap a few nuisance Bavarian beavers and ship them over? *Natürlich*. From 1996 to 1998 Schwab moved twenty-nine beavers to the confluence of Croatia's Mura and Drava Rivers, and another fifty-six to the Sava watershed.[14] The animals took. Croatia had been re-beavered — and it was only the first domino.

Over the next decade Schwab traipsed across Europe, scattering beavers in his wake like a Teutonic Johnny Appleseed. Two hundred and fifty-five beavers went into Romania's Olt and Iolamita Rivers; two hundred more splashed into national parks in Hungary; still others colonized Belgium and Spain and Bosnia. Schwab's methods were simple: pack the beavers into crates (his self-explanatory advice: "Use boxes made of metal, not wood"), stick them in a truck with a few apples, and drive. Only in the early years did border crossings prove a hassle. Sometimes beavers took relocation into their own paws: Schwab canceled plans to move them to Slovenia after the reestablished Croatian population crossed the border on its own.

Over time Schwab expanded his reach. In 2012 he arranged a shipment of fourteen beavers to Mongolia to cleanse and rewater the polluted Tuul River.[15] (That project, Schwab told me, was the first time Eurasian beaver reintroduction had been explicitly motivated by ecosystem services, rather than conservation for its own sake.) A few countries — Italy, Portugal, Albania — remain destitute, but Schwab expects beavers to find their way to those sovereigns without his aid. "There are also no beavers in the Vatican," he added with a raised eyebrow, "but they've got a garden with some water, so . . ."

I asked Schwab how many beavers he'd relocated in his lifetime, assuming he'd offer a ballpark estimate. Without hesitating, he replied, "Nine hundred and seventy-three."

"Which means you need — "

"Twenty-seven. And then I have my thousand full."

I sat back, a bit boggled. "So where will you send your last twenty-seven?"

"I'm hoping," he said, "that it will be Britain."

Spend enough time squelching through beaver country, and you'll start to feel like you're wandering through a recurrent dream. The backdrops change, but the visual motifs — the whittled stumps, the spongy ground, the scaffolding of downed wood — vary little; a beaver complex in Connecticut resembles one in Scotland resembles one in Wyoming. Beavers are inventive, to a point — Mike Callahan once showed me a picture of a dam built through the rusted cab of an abandoned pickup — but their works tend after a while to blend together: dam, lodge, pond, canals, repeat. We humans, for better or worse, tend to construct our habitats with a bit more whimsy. For proof, I show you the Bamff Estate.

Paul and Louise Ramsay, we found, lived in the sort of house that makes visiting Americans reflect on their own country's callowness. The Bamff Estate's main building was a gothic, imposing manor with cream-colored plaster walls, peaked slate roofs, towering stone chimneys, and conical towers from which I half expected to see Rapunzel lowering a golden braid. Some of its stonework dated to the sixteenth century. It could have housed a baron, or a boarding school, or a villainous witch. Two towering hedges stood sentinel over the garden, the topiary shabbily trimmed to form rearing animal heads.

"I love your red squirrels," I told Paul, with a nod to the hedges, during our tour of the grounds.

He winced. "Actually, they're meant to be unicorns," he said. "That one's just lost his horn to die-back."

The same frayed nobility pervaded the interior. The carpets and uphol-stery had worn a bit threadbare, and the plaster in the guest bathroom — complete with clawfoot tub, of course — was cracked. But somehow the palpable antiquity only heightened the place's grandeur. It possessed the magic of a fairy tale, the cockeyed darkness of a Roald Dahl novel. Generations of Ramsays peered down from oil paintings in the stairwells, pink-cheeked and robed in Victorian finery. The wood banister had been buffed to a deep shine by thousands of hands. A library-like perfume hung so thick we could see the dust motes. Books overflowed from cabinets, spilled from shelves, rose volcanically on countertops. Books with faded tattered spines, books with ornate gold inlays, books with names like *The Romance of Modern Locomotion*. Enos Mills's *In Beaver World*, I noticed, occupied a place of pride in the living room.

Though Paul and Louise had both grown up in Scotland, they spoke the clipped Queen's English that marked boarding school educations, pepper-ing conversation with urbane references to writers from Proust to Proulx. But they were hardly stuffy. At their dinner table, over chops and potatoes, we drank no small quantity of whiskey, traded no shortage of dirty beaver puns, and heard plenty of good-natured invective directed toward the farm-ers, fishermen, and officials who stood in the path of the rodent's return. Louise reserved special disdain for C. S. Lewis, who, in *The Chronicles of*

Narnia, had erroneously cast Mr. and Mrs. Beaver as piscivores, dealing an inadvertent blow to the castor cause. "In Britain, probably 75 percent of the populace reads Narnia in school, and goes their whole lives under the mistaken impression that beavers eat fish," she groaned. "It's made it quite difficult to talk to salmon fishermen."

The Ramsays stumbled into beaver reintroduction for a simple reason: No one else seemed inclined to tackle it. In 1992 the European Union passed a directive calling for member states to "study the desirability of re-introducing species," a command that prodded Scottish Natural Heritage, a governmental conservation body, to investigate the possibility.[16] The study crept along at a snail's pace, impeded by the anti-beaver lobbying of farmers and Narnia-addled fishermen. Paul Ramsay attended several official meetings, coming away certain that Scotland needed beavers, and equally certain that he would get old waiting for the government to release them. The Ramsays decided to take the future of Scotland's beavers into their own hands.

"Once you've been contaminated by an ecological education, you see," Paul explained to me, "there's no going back."

In 2002 they released their first pair of Norway-born beavers into an enclosure on the Bamff grounds. "The first lot were a disaster," Paul said: One was squashed by a falling tree, the other killed by a parasite. More beavers followed, though, from Poland and Bavaria — the latter courtesy, of course, of Gerhard Schwab. To the Ramsays' delight, these imports were a hardier bunch, and by 2005 they'd begun to breed and build. A ditch became a swamp; herons and woodpeckers filled the ponds and snags. The day we visited, we found the wetlands stamped with duck and deer tracks. Upon one log rested a juicy, slug-shaped otter *spraint*, or scat, packed with fish bones.

The Bamff Estate sits in the catchment of the Tay, Scotland's longest river system. In the early 2000s Tayside fishermen began encountering animals that looked an awful lot like beavers — the first wild ones to roam Scotland in four centuries. By 2012 the Tay population had exploded to around 150 beavers, and rumors swirled about its origins.[17] Some farmers and fishermen, perhaps understandably, pointed fingers at the nearby estate with the very public beaver colony. According to Paul Ramsay, though, the Tay's original colonists escaped from a nearby wildlife park.

And it's true that locals began reporting beaver sightings well before Bamff's beavers ever reproduced. "I can safely say that there were already beavers around the Tay before we had anything that was actually dispersing," Ramsay told me, a bit vaguely.

"So it's not true that the Tay population came from Bamff beavers," I said.

"It's not true," Ramsay confirmed. Then he paused and cocked his head. "Well, it's an exaggeration."

We had, at this point, come to the watergate that once formed the downstream end of Bamff's enclosure, a rather flimsy-looking barrier of posts and chickenwire. "I suppose it was a bit naive to think that this would be able to enclose a beaver," Ramsay said, sounding not particularly remorseful. Paul and Louise may not have actively released beavers, but they hadn't managed to keep them confined. "Where there's a will there's a way, and beavers, you see, have a great deal of will. The first exploratory jailbreak was in 2006, we believe. At this point we have no idea how many have escaped," he added with an amused shrug.

"The fact that your beavers dispersed into the countryside — did you ever face any consequences for that?" I asked.

"Oh, well, yes. I was arrested!" Paul exclaimed, with great merriment. "What happened was that a beaver gnawed the gatepost of a house on the way to Alyth. The police thought this was the moment they could finally put a stop to all this. And so I was rung up and asked to go to an interview in Perth, at the police station. At the end of the interview, when I thought we were all going to shake hands and get on with our business, the police officer said, 'I'm going to arrest you.' So I was arrested, and had my DNA taken, and my photographs, and fingerprints." He smiled faintly, relishing his tale. "Fortunately, just to be sure that my human rights were not going to be violated, they told me I was entitled to take a lawyer. My lawyer said, What you must do is write to them to explain what may have happened. Young beavers, you see, can travel enormous distances when they're dispersing. And so I wrote to the police, explaining that the fact that a beaver was *near* me didn't mean that it had *come* from me." Armed only with circumstantial evidence, the prosecution dropped the case.

As the Tayside beavers spread, they attracted attention — not all of it positive. Beavers clogged drainage ditches, flooded fields, and burrowed into riverbanks, destabilizing flood defenses and eroding farmland. In rain-drenched Scotland, water storage isn't always an asset; sometimes, it's a liability. "A lot of our land was bogs and marshes until four hundred or five hundred years ago, when floodbanks were built and it was reclaimed," Andrew Bauer, deputy policy director at Scotland's National Farmers Union, told me. "While some people may think it would be lovely to go back to that, from an agricultural point of view it would be disastrous. The economics of farming in Scotland means it doesn't take too many large floodbank breaches for even the largest agricultural business here to start struggling."

In the years after the beavers' arrival in the Tay, irate farmers gunned down some twenty hapless rodents.[18] "They were assassinated, you might say," Ramsay told me. "Murdered, even!" Landowners cleared hundreds of trees along riverbanks to discourage the animals, an ugly, counterproductive measure that one columnist called "environmental madness."[19] Former friends told Paul he belonged in prison; Louise was accosted by an angry farmer at a funeral. "The Ramsays are blinkered," one fisherman griped to the *Scottish Mail*.[20] "If a farmer broke wildlife laws, they'd rightly be held to account," Bauer told me. "Whereas these people have broken numerous rules and received no sanction."

The government was equally peeved. In 2010 Scottish Natural Heritage announced a plan to trap the Tay's beavers, house some at a zoo, and euthanize others. The department had, at long last, launched its own official beaver reintroduction, in a western region called Argyll, and seemed annoyed that the rogue Tay beavers hadn't followed proper bureaucratic channels. "You have to believe in upholding wildlife law or not," one executive sniped.[21] "I think this is quite simply professional jealousy," a beaver aficionado retorted to *The Guardian*.[22]

The Ramsays, outraged at the trapping scheme, launched a campaign to save the Tay's population, arguing that beavers were a reestablished native species that deserved legal protection. Along with other wildlife advocates, they formed a nonprofit, the Scottish Wild Beaver Group, and sought the aid of the global Beaver Believer community, enlisting Heidi Perryman of Martinez to coordinate their social media blitz. Rick Lanman, the California cancer geneticist turned natural historian, wrote beavers into the Tay's Wikipedia page. The campaign's turning point, in the end, was a wound self-inflicted by the government. In April 2011 the first and only beaver trapped from the Tay River, a female dubbed Erica, contracted an infection under the zoo's care and died.[23] The tragedy didn't go to waste. "We think the sad death of this young beaver emphasizes the cruelty and folly of the attempt to trap and remove these native animals," Louise told the BBC.

The government, chastened by the death and daunted by the Sisyphean challenge of eliminating the prolific population, suspended the trapping program. Erica died a martyr. The Tay's beavers remained.

———

In my experience the bitterest conservation battles aren't pitched against chemical companies or fossil fuel executives — they're waged between ostensible allies. Environmental campaigns tend to comprise unwieldy coalitions of groups and personalities who lie somewhere along a very wide spectrum from radicalism to pragmatism. Every American conservationist I've met wants to limit carbon emissions, protect public lands, and grow wildlife populations. How to get there, though, and how much to compromise: That's the rub. Everyone shares a goal; no one agrees on strategy.

One morning not long after arriving in Glasgow, Elise and I drove two hours through Scotland's glacier-scoured west, along the bonnie banks of Loch Lomond and past the trim town houses that line the Crinan Canal, stopping only for a breakfast of black pudding and tattie scones, to meet Oly Hemmings and Pete Creech in Argyll, the district that's home to the Scottish Beaver Trial: the authorized counterpart to the Tayside's scofflaw population. Hemmings and Creech co-direct the Heart of Argyll Wildlife Organisation, a group devoted to raising the region's profile. Although beavers had become Argyll's most famous wild residents, they were merely its most visible attractions. If you appreciate the subdued pleasures of mosses, lichens, and dragonflies, Creech said as we clambered into his car — Elise noted, approvingly, a beaver figurine perched on the dashboard — Argyll was the place for you. "It's an incredibly biodiverse area, and nobody knows about it."

Argyll is thickly forested and lightly peopled, which makes it an ideal place to release controversial creatures. Beginning in 2009 the Scottish Wildlife Trust and the Royal Zoological Society of Scotland, with the government's blessing, introduced sixteen Norwegian beavers to Argyll's icy lochs, granting them a five-year probationary status during which they could be trapped out if they misbehaved. Then the groups spent five years exhaustively monitoring their wards' impacts, collecting data on everything from stream flow to vegetation cover to beetle diversity.[24] In the end the research was most notable for what it *didn't* detect: Aside from a flooded path, locals' fears — clear-cut forests, inundated towns, blocked fish migrations — never transpired. The beavers passed the test.

Hemmings and Creech led us to one water body that beavers had remodeled: Loch Coille-Bharr, where the enterprising builders had submerged a popular walking trail. The Forestry Commission had ingeniously mediated the conflict by replacing the underwater trail section with a floating pontoon, which on this day was trafficked by a pair of elderly, binoculars-toting birders. From there we wandered along a boardwalk overlooking a shallow, muddy pond, its surface tousled by a cold breeze. Although the beavers had departed for another loch, their influence lingered: Skeletal birch and hazel picketed the pond, and tadpoles writhed in a canal.

"For us, this is a view that hasn't been seen in Scotland in four hundred years," said Hemmings, a warm, gentle woman with a lilting voice. "Standing deadwood in any other situation would be cleared away because it looks unsightly to us tidy humans." After the introduction of beavers, she added, "The dragonflies and damselflies started to hatch out, and the newts and frogs increased as well. In spring this place comes alive."

The transformation to Argyll's ecotourism industry might have been more profound. "People staying with us now put in the guest book, 'We've come three days to see the beavers,'" Creech, who also runs a bed-and-breakfast, told me. The mammals had become such reliable revenue generators that they'd won over their most stalwart detractor, a local basket weaver. "She probably had more reason to be frightened than anyone else," Creech added. "She grows willow."

From a demographic standpoint, though, the Argyll beavers were struggling. In contrast with their Tayside brethren, the official population had failed to increase: 71 percent of their wild-born kits had died.[25] The stagnant growth was largely by design. The government had chosen Argyll for the glacial topography — all deep vales and looming knaps, bounded by mountains and the sea — that formed a giant natural enclosure. The rugged terrain ensured that dispersing beavers wouldn't irk nearby landowners, but it had trapped the Argyll colonies in a bind. "The animals basically are not crossing the ridges and the valleys, so they're not finding each other," one British beaver advocate told me later. "There's as many beavers there as when the project started, and left to its own devices it would fail."

Partly for that reason, there wasn't historically much love lost between the trial's overseers and the Tayside crusaders, although Paul Ramsay told me he's become more keen on cooperating with the Scottish Wildlife Trust since the Tay beavers earned a reprieve in 2012. The Ramsays' upstart Scottish Wild Beaver Group feared that the minuscule Knapdale population could be easily squashed, while the trial's organizers harbored lingering indignation about the run-amok Tay population. "I think it's really, really important that people abide by the law and follow good practice," Susan Davies, director of conservation at the Scottish Wildlife Trust, told me. "With the Tayside, the fact that there was no agreed management approach in place just heightened tensions, and could have stymied the whole reintroduction process."

Rivalrous though they may be, the two populations — Argyll and Tayside, Tayside and Argyll — seem to me more symbiotic than competitive. The Tayside rebels, and the breathless media coverage they inspired, boosted the profile of Britain's beavers a hundredfold; they may have antagonized farmers, but they won thousands of Scots to the cause. Meanwhile, the government-sanctioned reintroduction demonstrated to anxious landowners that the return of beavers to Britain need not be a free-for-all. The laird and lady of the Bamff Estate may not always see eye-to-eye with the Scottish Wildlife Trust, but it takes all types to restore a species.

As I write these words, the future of Scotland's beavers remains murky. In the fall of 2016, Roseanna Cunningham, Scotland's environment secretary, announced that she was "minded" to allow beavers to remain in both Argyll and Tayside, a cagey phrasing that authorized their presence without formally protecting them.[26] Until the Scottish parliament ratifies protection, there's nothing stopping farmers from blasting away at every beaver that materializes in their fields. Although the Scottish Wildlife Trust received permission to augment the Argyll population with twenty-eight new beavers,[27] don't expect to see the group establishing new colonies anytime soon. "We believe that potentially new release sites would be required to get the strongest, most sustainable population," Davies told me. "But politically, that isn't the place where we're at."

A bedrock tenet of conservation biology is that healthy wildlife requires connectivity. Isolated islands of animals, cut off from adjacent populations by towns or roads, are more vulnerable to blinking out than groups that can mingle and mate with their fellows. That's as true for beavers as for bears or bison. When we visited the Bamff Estate, the Ramsays had just received encouraging news on that front: Argyll beavers had finally dispersed east, and the metastasizing Tayside unit had infiltrated Loch Tay. "It's getting very close now," Louise told me. Only a handful of miles separated Scotland's pioneering beaver populations. Connectivity, and the brighter demographic prospects that came with it, seemed inevitable. With government aid or without it, beavers have returned to Scotland for good.

———

Dumping beavers in the Scottish Highlands, where the terrain is jagged and the homesteads sparse, is challenge enough. Take a nine-hour train ride south to England, though, and you'll find yourself encircled by a clutter of farms and towns dense enough to make Scotland look like the Alaskan bush. And yet West Devon, a pastoral district perched on England's southwesternmost tongue, might hold the key to the future of British beavers.

Driving the Devon countryside is not recommended for the claustrophobic. Bumpy roads no wider than sidewalks meander across the borough, bracketed by looming, impenetrable hedgerows that can be neither peered over nor seen through, an effect that can make visiting Americans feel like lab rats attempting to navigate a maze of foliage. Like the inept Yank I am, I'd been unable to drive the manual transmission the car hire provided me in Exeter; the only automatic in town was a boat-sized SUV that made as much sense in these hedge-bound roads as trying to pilot an icebreaker through the canals of Venice. Garlands gathered on our side mirrors as we pinged off the greenery, occasionally reversing to make way for the odd horse trudging down the lane.

Where the hedges broke, we saw sheep, sheep, and more sheep: fluffy, indolent, staring. A laminated poster hung along a walking path brooked no doubt about which animal owned the fields. "There was a severe case of sheep-worrying on Friday 14th Oct, when a BLACK DOG, out of control, chased the in-lamb ewes to the point of great distress," the sign admonished. And below that, an ominous addendum: "BE WARNED that any dog found worrying the sheep can be legally shot dead." We weren't sure whether human sheep-worriers also faced capital punishment, but we resolved to give the ruminants a wide berth.

Only at night did the agrarian landscape permit the incursion of wildness. Driving back from the local pub, our SUV's headlights illuminated rabbits, mice, voles, and stoats. In the morning, the only sign of these nocturnal travelers was the sporadic *smeuse* — a Sussex dialect noun delightfully captured by Robert Macfarlane as "the gap in the base of a hedge made by the regular passage of a small animal."[28]

West Devon looked, in short, too domestic to host an ambitious species reintroduction. But appearances mislead: The region boasts England's only wild beaver population, a community of twenty rodents flourishing in the River Otter. (Yes, the mammal dissonance is confusing.) Although reports began circulating as early as 2007, it wasn't until 2014 that photographs confirmed the presence of the animals[29] — fugitives from a private collection, most likely. The river's deep channels and burrow-friendly banks allowed the creatures to flourish for years without betraying themselves by building dams and lodges.

As in Scotland, the English government was none too keen on allowing the escapees to remain. But locals enjoyed having them around, and business improved at riverside pubs as patrons flocked to catch their first-ever glimpse of a wild castorid. In 2015 the Department for Environment, Food, and Rural Affairs (DEFRA) live-trapped the entire population and tested their genes to ensure the beavers were *Castor fiber*, rather than their loathsome North American cousins. Finding the beavers both European and disease-free, DEFRA returned them to the River Otter.[30] But their freedom was only provisional. In 2020, the agency vowed, it would make a final decision. Cause too much trouble in the meantime, DEFRA promised, and the beavers could be removed as abruptly as they'd appeared.[31]

On a rare sunny afternoon, Elise and I coaxed our monstrous car through the countryside to meet an expert witness testifying in the beaver's defense. We met Alan Puttock at a pullout bordered by, yes, a hedgerow. Puttock, a University of Exeter hydrologist, was a scientist straight out of *Big Bang Theory* casting: brilliant, sweet-tempered, and endearingly awkward. We'd come to join Puttock on a beaver tour he was leading for a farming advocacy group. As the convoy approached, he laughed nervously.

"I'm meant to be in a lab somewhere," he admitted. "But to be honest, no one's going to read my papers. They're really boring — they've got loads of standard deviations and stuff in them." The only way to win converts was to brave a field trip. Puttock put on his game face.

After the group assembled, Puttock led us down sloping fallow fields toward a straggly wood. A stunted black-and-white pony watched us pass, chewing stolidly. The sun had baked the spongy earth a bit, but I regretted being the only attendee who'd forgotten wellies. Before long we came to a fenced paddock, its interior thick with brush. On most farms electric fences keep nibbling creatures out. But this one had been built to keep animals *in*: a family of four beavers, descended from a pair released into the enclosure in 2011 by the Devon Wildlife Trust. While their River Otter cousins roamed free, this colony dwelled in captivity on a seven-acre parcel of willow scrub and grassland bisected by a gurgling stream. But if these beavers — along with Puttock and his colleagues — did their jobs, they could jump-start reintroduction throughout Britain.

Within the paddock we found turbid ponds, snaky beaver-dug canals, and a potpourri of fragrant, gnawed chips piled beneath precarious trees. I had squished through a thousand complexes like it. But to most of the dozen or so farming advocates, I realized, this scene was utterly novel — a bewildering, breathtaking phenomenon missing from these lands for centuries. Despite Puttock's best attempts to herd his guests, they scattered through the site like cattle turned out to pasture, their colorful raincoats flickering through the oak. (In West Devon you wear your raincoat even when it's not raining, because you never know.) They stepped, cautiously and then with growing delight, onto the primary dam, flexing their knees to test its integrity. Puttock chased after them, answering the usual questions: Why do they build dams? Which trees do they prefer? And, inevitably: Do they eat fish?

"What's quite cool is that there's also a bachelor lodge in here," Puttock said, gesturing to a low muddy mound. "So when the female got pregnant, she chucked out the male, and he was made to live in there for a while."

"My daughter's husband would have quite liked that, actually," said a short woman in a blue slicker.

"Do you know how much water they're trapping on the land here?" asked a gangly man.

Puttock nodded vigorously. "It changes with rain events, but we've had up to one million liters of water being stored in the ponds at any one time," he said. "They're kind of creating water out of . . . nowhere."

Although Alan Puttock's papers are, as he says, full of loads of standard deviations, I can attest that they're anything but boring. If you love beavers, his recent work in *Science of the Total Environment* might as well be a Michael Crichton novel. In 2014 he and his Exeter colleagues set up two sets of instruments — one just upstream of the enclosure, one just downstream — that essentially counted every drop of water that entered and exited the site, passing through thirteen beaver dams along the way. A satellite link beamed every byte of data back to the office in Exeter, the connection powered by a small on-site solar panel.[32]

The result was one of the clearest snapshots of beaver influence ever taken. As you'd expect, the Exeter team found that the compound stored water and filtered out nitrogen, phosphorus, and sediment. The number of gelatinous frogspawn clumps at the site skyrocketed from just ten before beaver reintroduction to five-hundred and eighty several years later, and water beetle diversity climbed from eight species to twenty-six.[33] What most impressed the researchers, however, was the site's ability to buffer floods. During downpours, a stunning 30 percent of the water that entered the enclosure never left it. Instead runoff vanished into ponds, dispersed laterally into wetlands, or sank into the ground.[34] The complex functioned like a towel held under a faucet, absorbing tapwater before it could reach the drain. "If this was a small straight channel, with no ponds, that wouldn't happen," Puttock told the group.

What wildfires are to the United States, floods are to the United Kingdom: chronic natural disasters exacerbated by shortsighted management. Given its sodden climate, Britain will always be flood-prone; its unremitting agricultural development seals its fate. Canalized farmland funnels rainfall downstream instead of permitting it to percolate into fields and woodlands, worsening winter floods that tear through towns and cities. Torrents in northern England in 2015, for instance, cost the nation more than five billion pounds — nearly seven billion dollars.[35]

If you live beside a beaver pond, you likely — and rightly — consider your resident rodents to be flood *risks*. But look beyond your backyard, and you'll see that beavers have a well-documented ability to abate, absorb, and attenuate floods at a vast scale. Consider the Ourthe Orientale, a river in southern Belgium that, between 1978 and 2003, was swollen by high flows every three and a half years or so. After *Castor fiber* returned – aided by Gerhard Schwab, who turned ninety-seven beavers loose in the region – floods became significantly rarer, occurring only once every 5.6 years. The river's roller-coaster-like hydrograph turned gentle, as chains of dams and ponds swallowed floods and lowered "discharge peaks."[36]

Given those results, you might think farmers would be over the moon about beavers' potential. After Alan Puttock finished his spiel, I piped up to ask the farming advocates how their constituents might react to a full-fledged re-beavering. There was a silence, followed by a collective giggle. I saw them cast sidelong glances, asking with their eyes, *Who wants to answer this ignorant Yank?*

Finally, a sturdy, clean-shaven man spoke. "I think they'd be deeply skeptical," he said, to nods. "We work for a charity whose farmers are typically engaged with wildlife. But even then, they'd all have commercial concerns about the level of flooding across their land. When I've talked to people, they've just said, 'I'm going to end up with a lot of pointy stubs in the ground, and not much else.'"

So far, thankfully, no one in West Devon has ended up with nothing but pointy stubs. For now the River Otter's beaver population is small enough, and the habitat prime enough, that the animals can generally make a living without building dams. When I spoke to Mark Elliott, beaver project leader for the Devon Wildlife Trust, he told me that locals had only encountered a few problems with chewed trees – easily solved with a coat of abrasive paint, which discourages gnawing – and a single bout of dam-related flooding, addressed with a flow device.

Still, Elliott knows that England's beavers are coasting through a finite honeymoon period. At some point the rodents will spread into more marginal areas, building dams and impounding water as they go. The novelty will wear off. Weathering the inevitable conflicts will require problem-solving strategies. That might mean employing a British version of Mike Callahan, flow device kingpin of Massachusetts, or training legions of beaver-smart volunteers, Gerhard Schwab's approach to ameliorating beaver problems in Bavaria. Whatever the model, it would require new funding. Among other sources, the Devon Wildlife Trust had been paying for its beaver program by selling gnawed woodchips online ("in a protective box with card of authenticity"), a clever idea that didn't strike me as the most sustainable revenue stream.[37]

That's where the University of Exeter's research comes in. If Puttock and his colleagues could slap a dollar value on flood prevention, Elliott said, it might prove, even to skeptical farmers, that beavers' benefits outweigh their costs, generating governmental support — and funding. Accomplish that analysis by the 2020 deadline, and the pipedream of widespread reintroduction could become reality. In fact, it might happen anyway: Several months after my visit, British environment secretary Michael Gove announced that beavers would be returned to the Forest of Dean, leading enthusiastic local councilors to rejoice at the prospect of flood alleviation.[38] Since 2005 supporters have also pushed hard for reintroduction in Wales, where the long-absent creatures are known as *afancod*. "It's all very well to put more beavers behind fences," Elliott said. "But we want them back out into the landscape where they can do the most good."

Although flooding on our side of the Atlantic is not quite as epidemic as it is in Britain, we North Americans could learn a thing or two from the Exeter lads. Our most famously flood-prone basin is the Mississippi, where two centuries of levee building, development, and channelization have principally served to piss the Big Muddy off. "Once more war is on between the mighty old dragon that is the Mississippi River and his ancient enemy, man," intoned *The New York Times* in 1927, when floods inundated twenty-seven thousand square miles and killed up to a thousand people.[39] The tighter we wrap our fingers around the Mississippi, the more fiercely it writhes in our grasp: In 2017 the river reached near-record crests, inflicting tens of millions of dollars in property damage and putting ten million midwesterners under flood warnings.[40]

For all the shots we've fired during our war with the mighty old dragon, the most injurious might have been the first: the trapping of the Mississippi's beavers. Writing in the journal *Restoration Ecology* in 1995, two years after a spate of flooding cost the Midwest sixteen billion dollars, Donald Hey and Nancy Philippi estimated that beaver ponds once covered up to fifty-one million acres of the Upper Mississippi Basin — more than 10 percent of its total area. When those ponds dried up, much of the Midwest's absorptive capacity vanished. "At a depth of three feet," Hey and Philippi wrote, "the original ponded area could have stored more than three floods the size of the 1993 event."[41] Paving over wetlands and reflexively killing problematic beavers are ecological injuries masquerading as progress — a privileging of the built environment over natural infrastructure. In the end, both suffer.

———

It is easy for North American wildlife-lovers to take for granted our good fortune. Sure, we lost the passenger pigeon and the Carolina parakeet and the sea mink, but we managed to pull most charismatic species — bison, grizzlies, even deer — back from extinction's precipice. In many cases returning wildness to North America has simply required staying out of the way. Although some critters — the California condor, the black-footed ferret — demand active rehabilitation, many others require little more than protection against mass slaughter. The cougar, for instance, whose numbers rebounded from two thousand to thirty thousand by the early 1990s, didn't require reintroductions, captive breeding, or even much dedicated habitat preservation. We just eased up on the guns and the traps and the poisons. The species' own resilience prevented us from making an irreversible mistake.

Britain's fauna suffered a far worse fate. Although North America's indigenous people cultivated and deliberately set fire to millions of acres, the British Isles were subjected to far more intensive burning, clearing, and grazing, and for far longer. The country's agricultural systems are so pervasive and so ancient that they're practically invisible, unless you know what you're looking for. One typically damp day Elise and I drove through Dartmoor National Park, the nearly treeless moor whose eerie wastes inspired *Hound of the Baskervilles*. We were charmed by the heather and rocky tors, features that felt quintessentially British in their sober gray-scale — until we learned that the land's baldness was the depressing legacy of millennia of cutting, fires, and livestock. Sheep still cloak every inch of Dartmoor, while large wild animals are woefully absent from both national park and Britain at large. Wolves? Gone from the isles since 1621. Lynx? Wiped out in the sixth century. Moose? Extinct for four thousand years. In Africa safari-goers can behold the Big Five mammals, a group composed of elephants, lions, rhinos, buffalo, and leopards. At Dartmoor, a sign at the visitor center informed us, we'd be treated to the spectacle of the Little Five, a list that includes a beetle, a butterfly, and the ash black slug. I appreciate a fine gastropod as much as the next guy, but the Serengeti it ain't.

"By the time you're down to saving bees and ants, you know you're buggered," an English conservationist named Derek Gow told me. "We've exterminated everything that counts. You're looking at a country that's in virtually a complete state of ecological collapse. When you look out that window and it's green, well, there's nothing much actually living in the green, unless it can live in a bloody hedge."

More than most, Gow knows what it means to sweat the small stuff. On yet another wet, shrouded day, we met Gow at Upcott Grange, a sprawling farm in Devon's hinterlands. Upcott Grange might be the only farm in England whose focal building is a wooden shed labeled VOLE ROOM. When we followed Gow into the Vole Room, we found it crammed with metal-framed cabinets and sliding plastic bins, like an elaborate filing system. The room was redolent of sawdust and fur. In each plastic bin resided, of course, a vole — an adorable ball of charcoal-and-chocolate fluff with glittery black eyes and a faceful of whiskers. One worker was busy cramming a vole into an empty Pringles tube for transport; the vole clung to the lip, resisting internment.

Water voles — semi-aquatic rodents that burrow into streambanks — are Britain's fastest-declining mammal, pushed toward oblivion by habitat loss and non-native predators. Gow has spent much of his odd career captive-breeding and reintroducing the fuzzballs, which, in the mid-1990s, got him thinking about larger rodents. "As soon as you look at the ecology of little aquatic animals like water voles, which need open, sunny, complex pool systems, you realize there must have been something making those systems," he said as we proceeded from the vole room to a drafty office for a cup of tea. "That starts you thinking, and you realize it must have been beavers."

Gow is a hard guy to pigeonhole: He's a farmer, of sorts, with an education in agriculture and livestock on his land. His real passion is husbanding odd animals: When you Google his name, the results are dominated by articles chronicling his ill-fated attempt to raise "Nazi super cows," a freakishly muscular breed first cultivated by German zoologists; Gow had to put down his Aryan cattle after they tried to murder his staff.[42] Although the Nazi cows are no more, Upcott Grange still hosts a captive menagerie that draws photographers and filmmakers from around the world: marten, mink, otters, deer, owls, boar, and many other beasties, including beavers. In the self-serious world of conservation, Gow is the rare cad who recognizes the comedy inherent in obsessing over rodents.

"Beavers seem for some strange reason to dominate my life," he said with a mordant sigh. "At my birthday I was having another fucking argument about beavers with the government, in another village hall, and a friend sent a message saying, Happy birthday, Derek, what're you doing? And I said, I'm in another fucking village hall, having another fucking argument about beavers. And he said, Well, having known you for twenty-five years, you're probably happier doing that than you'd be doing anything else. And I said, Well, yeah, it actually *is* fun."

Gow's animal husbandry skills play a critical role in the return of Britain's beavers. Any castorid that enters the United Kingdom from Germany, Norway, or elsewhere must be quarantined for six months, to ensure the new arrival doesn't carry rabies or a nasty liver-attacking tapeworm called *Echinococcus*.[43] Most beavers endure that quarantine period at Upcott Grange. Among the rodent immigrants Gow cared for were the Ramsays' beavers before they were released into Bamff, the beavers that occupy the enclosure in West Devon, and the Scottish Beaver Trial releases. Along the way, he has become a formidable beaver advocate who seems to particularly enjoy provoking the landed gentry.

"Britain is still a heavily socially stratified society," he told us. "There have been conversations in the drawing rooms of the great stately homes where some buff tells some other buff that beavers are like rats or gray squirrels, they'll be everywhere and do tremendous damage, and his father never would have given in to the despicable greenies. And you'll never even get to meet most of them making these decisions — which, I have to say, I don't consider any kind of social loss."

Nonetheless, Gow has a surprisingly sunny outlook on reintroduction's prospects. Although I found it hard to imagine that Britain was ever anything other than gloomy, Elise and I had actually arrived just after the driest winter in twenty years. Ping-Ponged between drought and flood, Gow said, Brits were "really starting to think about whether we can have a sustainable relationship with water in this country without this animal." Pro-beaver articles were popping up in the BBC, *The Guardian*, and the London *Times*, like long-suppressed samizdat finally seeing the light of day. "Some of the old organizations who've been very opposed to beaver reintroduction, like the farmers' unions and the hunters' associations — we're having some very interesting conversations with them now," Gow said.

If beaver restoration succeeds, the prodigal mammals may nudge open the door for further faunal comebacks. In the United States we're lucky to adjoin Canada, a wooded and wild refuge for many persecuted species; long before biologists returned wolves to Yellowstone, for instance, packs were naturally venturing into Montana from Alberta. Britain, however, has no connected neighbor to bail it out, meaning that bringing back its wildlife requires more active intervention — a process often referred to as *rewilding*. It's a contested word, deployed by various factions to describe the resumption of natural processes, the restoration of extinct ecosystems, or simply a greater intimacy with hunting and gathering. In his book *Feral*, George Monbiot defines it as a recognition that "nature consists not just of a collection of species but also of their ever-shifting relationships with each other and with the physical environment."[44] Fundamental to Monbiot's version of rewilding is allowing ecosystems to shape their own destinies, rather than micromanaging them. By that standard, beavers, which drive ecological processes like no other creature and bend to no person's will, are rewilding's poster species.

Reintroductions in Britain have proceeded tentatively, a reflection, perhaps, of a country known for its parochialism. In 1975 conservationists relocated sea eagles — hook-beaked raptors whose enormous wingspans earned them the nickname "flying barn doors" — to Scotland. Today around a hundred nesting pairs swoop among the salt-swept crags, and occasionally carry off a lamb. Red kites, osprey, corncrakes, and great bustards followed. Pool frogs were reintroduced in 2005, and a pair of insects, the large blue butterfly and the short-haired bumblebee, now float over hill and dale.[45] Beavers remain the only mammal yet reintroduced, though proponents hold out hope for lynx, moose, bison, and even, someday, wolves.

Could beavers serve as an exemplar, the species that teaches Brits that it's possible, and even worthwhile, to coexist with disruptive animals — a gateway mammal, if you will? When I posed the hypothesis to the Devon Wildlife Trust's Mark Elliott, he rejected it. "I've got a lot of sympathy for the restoration of some of our former species, but I don't see the beavers as a trailblazer for that. The impacts it has are unique," he told me. "We don't want the farmers and landowners we're working with to think that we're about to rewild their farms. We're not talking about lynx, and we're not talking about wolves or bears or anything else."

If rewilding the close confines of Devon is a bridge too far, the Scottish Highlands might offer more hope. The Scottish Wildlife Trust included lynx reintroduction in its list of fifty things that it hopes will happen in the next fifty years, and a group called Lynx UK Trust has proposed releasing as many as 250 of the tuft-eared cats between Loch Lomond and the West Highlands.[46] If farmers someday accept a re-lynxing, Susan Davies of the Scottish Wildlife Trust told me, it will be because beavers first earned their trust. "Lynx are doable, if we do it in the right way," Davies said. "But we recognize that we have to take it quite slowly, and address the management issues with beavers first." Although beavers deserve to return to Britain on their own merits, they're also rewilding's most crucial test to date — and, perhaps, its vanguard.

———

After tea, Gow led us into a dim, hangar-sized barn whose walls were lined with aluminum-sided enclosures, like pigsties. Wind hissed through the gaps in the slats, and our feet echoed on the concrete floor. A collie trotted along at our heels. Gow scratched him between the ears. "Wouldn't you like to be a beaver hound someday, Rexie?" he cooed. Gow opened the door to one of the pens and squeezed in, beckoning us to follow. The ground was strewn with a layer of straw and the odd apple; the smell recalled horses. He lifted up a metal sheet that served as the roof of a crude domicile, revealing a chocolate-colored Eurasian beaver within — the closest I'd ever been to *Castor fiber*.

This Bavarian beauty, Gow said, was a female, the latest Gerhard Schwab special to do six months of hard time at Upcott Grange. Her quarantine was nearly up: Next month she and a male companion would be released into a five-acre enclosure in Cornwall, a reintroduction crowd-funded by legions of online beaver boosters.[47] The nearest town, Ladock, had been doused by floods in 2014; the University of Exeter team would study the new releases to see whether the same flood attenuation benefits they'd documented in Devon would hold true in Cornwall.

Most biologists only recognize subtle differences between Eurasian and North American beavers — the latter have slightly larger skulls and broader tails, for instance. Gow, however, swore you could tell the species apart at a glance: *Castor fiber* is a bit thicker-set, he said, and waddles with a different gait. To my untrained eye, this beaver would have been perfectly at home in suburban New York. I crouched in front of the female, flashing my camera in her face. The beaver, vexed by the noisome intruder, growled, a low guttural note of warning. Like the blithe tourist I was, I continued shooting, dimly aware that I'd irritated one of the locals but nonetheless determined to obtain my photographic souvenirs.

"They're by and large pretty genial animals," Gow cautioned, "but this one's pissed off because — "

On cue the beaver lunged at me, agile as a cobra, and snapped, incisors closing briefly around my ankle. I yelped in surprise and stumbled backward, skidding on loose straw. Elise, watching from outside the pen, laughed at my dismay. Gow chuckled, too. Safely free of the enclosure, I rolled up my pant leg. No blood, no mark. It had just been a nip of admonishment, I realized, intended to deter rather than maim. "It's amazing how bloody high they can jump," Gow said. "One jumped up once and took a hole out of my T-shirt."

Having vanquished her foe, the Cornwall-bound female returned to her corner to triumphantly devour an apple. I watched from a safe distance, grateful she'd shown my ankle mercy — and pleased, too, that she hadn't acclimated to her human captors. She was a creature wary and willful, untamed and headstrong. A foot soldier in the rewilding.

CHAPTER TEN

Let the Rodent Do the Work

Like many cities that sprawl across the floodplains of the Mountain West, Logan, Utah, a municipality of fifty thousand tucked in the state's northeast corner, is a Jekyll-and-Hyde community. On its residential side children dance through sprinklers, locally owned markets perch on maple-shaded corners, and couples stroll up Canyon Drive to watch the Cache Valley turn lilac at sunset. In contrast with this bucolic slice of Americana is Logan's commercial drag, an unwalkable mallscape of casual dining chains and tire retailers and Walmarts. Note the plural: Since one big-box behemoth was not enough to satisfy Loganites, the Walton family's minions erected a second in 2006, constructing the new outlet alongside a small stream called Spring Creek. Several years after the store opened, beavers colonized its parking lot.

The occupation triggered the usual cycle: The city forced Walmart to hire a trapper. The beavers came back. Walmart hired a trapper. The beavers came back. The loop might have repeated ad infinitum were it not for Walmart's manager, who'd come to appreciate his rodent neighbors and felt dismayed at their repeated demise. The manager's concerns reached the ear of the local watershed council, which in turn approached Utah State University's resident beaver experts, Joe Wheaton and Nick Bouwes. It fell to Wheaton, a geomorphologist, and Bouwes, an ecologist, to break the cycle.

One scorching July afternoon Wheaton drove me to Logan's second Walmart to show me the leading edge of his plan to permit beaver-built chaos back into American life. Although Spring Creek was appealing enough — fringed by willow and cattail, shaded by a gallery of magisterial cottonwoods, slowed by a few small beaver dams — it seemed like a dubious place to start a family. Beyond one bank, a cluster of automotive shops performed serial oil changes on a parade of SUVs. Beyond the other, across a parking lot larger than some national monuments, shoppers filed out automatic doors, handcarts overflowing with grills and microwaves and flatscreens. To Wheaton, though, the urban setting only made the watering hole more extraordinary. "The biodiversity in here, the amount of wildlife and waterfowl and trout we see in this little nothing tributary?" Wheaton gushed from beneath a straw sunhat. "It's just crazy."

Even so, Logan's managers were reluctant to permit the beavers to remain. "The city had a very traditional risk-averse approach," Wheaton told me. "What do beaver do? They build dams — and they don't build their dams to code." Engineers feared the dams would fail, clog nearby culverts with debris, and wash out roads. "Any public works guy you talk to, that's going to be their view."

Over the course of a year, Wheaton and his Utah State team helped the city craft a management plan that addressed Logan's concerns without necessitating endless trapping. Worried about toppling cottonwoods crushing parked cars? We can smear trees with abrasive sand-and-latex paint to deter chewing. Freaked out about flooding? We can install a flow device. Heavy rains threatening to blow the dam? Use a backhoe to notch the dams or tear them out altogether. Only if an array of coexistence measures failed did trapping become an option — and even then, the plan, implemented in 2015, prioritized relocation over killing.[1]

"Instead of preemptively going to a conservative response now, let's put in some triggers," Wheaton said as we leaned against the chain-link fence that surrounded the wetland. "If we can figure out a way to eke out some gains in places this screwed up by allowing a damn rodent to do the work, that's just common sense."

Let the rodent do the work is one of Wheaton's favorite maxims. Like Brock Dolman, he is a geyser of colorful catchphrases and evocative analogies. Rivers that spill onto their floodplains are like teenagers attending keg parties; hack restoration jobs are akin to failed surgeries; heavy-handed stream modifications recall storming into Iraq and deposing Saddam Hussein without filling the resultant power vacuum. (That one might be a stretch.)

He levels profane, whip-smart stem-winders against everything from the exploitative economics of big-box stores to the government's response to Hurricane Katrina. "I'm like a windup toy that you just need to turn off sometimes," Wheaton warned me soon after we met.

Few topics rile Wheaton quite like his fellow stream restoration practitioners — particularly the class of consultants and engineers unshakably fixated on what Wheaton calls "Tonka Toys." Frontloaders, backhoes, and bulldozers are the standard tools of the restoration trade, the hulking yellow machines that contractors around the world deploy to gouge out pools, scour meanders, build riffles, and generally attempt to re-create natural stream features. While Wheaton acknowledges that some disastrously degraded waterways call for invasive surgery, he's adamant that the exorbitant expense of traditional restoration projects — $500,000 per mile is not an uncommon price tag — doesn't scale. One telling Environmental Protection Agency report found that, if you laid all American streams deemed in "poor biological condition" end-to-end, the watery path would stretch to the moon and back.[2] By my back-of-the-envelope calculations, attempting to repair all those streams would eat up the EPA's entire budget — for six consecutive years.

Restoring streams on the cheap, though, doesn't always come naturally to the restoration industry. "You convince your clients, you convince the regulatory agencies, you convince the world that only an engineer can do this stuff — it's so technical, it's so complicated," Wheaton groused as we drove back toward his office on the Utah State campus. "But engineers are focusing on the wrong questions."

Wheaton, a California native, would know: He was once an engineer. His first job was with a boutique firm in the Bay Area, controlling erosion at small-batch vineyards for, as he put it, "Dot-com bubble idiots from Silicon Valley who made way too much money overnight." He stumbled into geomorphology and fell in love with research, relishing the joy of chasing his own curiosities rather than kowtowing to clients. When he landed at Utah State University in 2009, he met Nick Bouwes, one of the architects of Oregon's Bridge Creek Beaver Dam Analogue project. The two became fast friends and collaborators — and, soon, beaver acolytes. "From the second we went out to Bridge Creek and I saw what this ecosystem engineer was achieving, I was sold," Wheaton told me.

In the years since his conversion to Beaver Believer, Wheaton has further refined Bridge Creek's beaver-based approach to restoration — a strategy he characterizes with another catchphrase, *cheap and cheerful*. To Wheaton's mind, the proving ground of Bridge Creek, where dam analogues and their rodent maintainers doubled fish survival at a fraction of the cost of conventional restoration, is a waypoint en route to a thorough re-beavering of the American West. Increasingly, he's focused on the most difficult projects in the most marginal habitats: restoring beavers, and the riverine processes they facilitate, to airport drainage ditches, blown-out creeks overrun with invasive reeds, and, yes, Walmart parking lots. The more debilitated the stream, the better. "I want to know where the breaking points are," he told me, "where this absolutely can't make sense."

The colossus who bestrides the narrow world of stream restoration is a swashbuckling hydrologist named Dave Rosgen, a charismatic figure known widely as the Restoration Cowboy. Since the early 1990s thousands of wide-eyed natural resource professionals have watched Rosgen, resplendent in white hat and fist-sized belt buckle, sermonize at one of his legendary fifteen-hundred-dollars-a-head workshops. Luna Leopold, the Cowboy's mentor, once called his pupil "the most outstanding practitioner of small-river restoration in the United States."[3] Some agencies consider Rosgen training a requisite for their contractors.

Rosgen's restoration strategy, called Natural Channel Design, amounts to a high-stakes wager that enlightened engineers can create waterways that look like natural streams — and stay that way. Its precepts are complex, but, at bottom, Natural Channel Design offers a recipe for transforming deteriorating systems into stable ones. Although the Restoration Cowboy strives to replicate natural riverine features, his tools — bulky rock and log structures called J-hooks, cross-vanes, W-weirs, and revetments, engineered to remain in place come hell or high water — are decidedly heavy-handed.[4] Rosgen famously transformed his signature site, Colorado's San Juan River, from a braided channel to a meandering one by getting behind the wheel of an earthmover and scraping out a new bed altogether.[5]

Rosgen's techniques have been applied to countless streams, and no wonder. For better and often worse, we are a nation of floodplain dwellers and farmers, drawn to river valleys yet intolerant of riparian anarchy. Natural Channel Design asserts that river professionals can have their cake and eat it, too — that natural-seeming streams can also be well-behaved ones, that carving out viable habitat doesn't require jeopardizing infrastructure. "There's people in the floodplain and you have to manage for them — you can't let the river do whatever it wants anymore," Kathy Raper, surface water manager for Wyoming's Sublette County Conservation District, told me. Raper has taken two Rosgen courses, and once recruited the Restoration Cowboy to hold a workshop in the Cowboy State. "If you let the river do what it wants, you're going to lose roads and subdivisions."

Rosgen's critics are nearly as numerous as his acolytes. Chief among the charges against Natural Channel Design is that its preoccupation with stability ignores how dynamic natural rivers can be. The single-thread channel winding like a giant *S* across the floodplain may be our platonic stream ideal, but it's also an artifact of logging, grazing, trapping, and other land-use changes. "The current condition of single gravel-bedded channels," wrote the geologists Robert Walter and Dorothy Merritts in 2008, ". . . is in stark contrast to the pre-settlement condition of swampy meadows and shallow anabranching streams."[6] Although Walter and Merritts were describing creeks in the mid-Atlantic, similar tumult prevailed in many western valleys. Pandemonium isn't convenient, but often it's more natural than stability.

The Restoration Cowboy's detractors — a group of geomorphologists so vociferous that Indiana University geographer Rebecca Lave dubbed the feud "The Rosgen Wars" — tend to congregate under the banner of "process-based" restoration.[7] If the Rosgenauts are concerned with how restored rivers *look*, the process tribe cares more about how they *act* — how often they spill onto their floodplain, how much sediment they convey, how nutrients cycle through them. To those who preach the process-based gospel, the best thing you can do for a river is permit it to obey its instincts — in short, to make a mess. You can probably understand why that message might frighten a flood control district, or perplex a restoration consultant trying to fix a degraded stream. "Traditional stream restoration is much more 'set it and forget it,'" Nick Weber told me when I visited Bridge Creek, "where you're trying to create a static channel that neither erodes nor aggrades." Beavers do many things well, but stasis isn't one of them.

Scientists enjoy nothing more than a good academic spat, and we journalists love to fan the flames of controversy. The camps' legitimate disagreements have been somewhat overblown: A 2016 paper on which both Rosgen and Joe Wheaton served as authors, for instance, noted that the lines between how form- and process-based systems classified streams were "not necessarily clear."[8] When I contacted Rosgen to solicit his thoughts on beavers, he replied with a lengthy email that, for the most part, lauded their role in restoration. When he's confronted with a hopelessly incised river, he wrote, he'll often dig a new channel across the floodplain and convert the abandoned trench to a series of oxbow ponds, which attract beavers, support fish, and raise water tables. It was beaver mimicry by Tonka Toy.

Although Joe Wheaton is the furthest thing from a Natural Channel Design proponent, he nonetheless admires the Restoration Cowboy. "That guy's done more to promote this obscure discipline called geomorphology than any geomorphologist," Wheaton told me. "Rosgen gets beat up for giving people recipes that they could follow in any situation, but that was the thing he got most right." Wheaton aspired to add a new batch of beaver-influenced formulas to the stream restoration cookbook. "If you're making cookies, Rosgen gave you the recipe for a Chips Ahoy," he said. "Whereas you could make this gooey, wonderful, delicious cookie with all these variations, with M&M's and sprinkles and nuts." There's a time and a place for bolted-down structures, particularly near roads and buildings — but there are still more places, Wheaton told me, "where we can get away with recipes that taste better and are better for the ecosystem."

So how do you craft a recipe for chaos? After our Walmart stop, Wheaton drove me an hour north of Logan to meet his partner in cookie baking, a rancher named Jay Wilde. Wilde, who sported a scrub-brush mustache and carried a tin of Copenhagen in the breast pocket of his pearl-snapped shirt, called himself an "old hayseed," but his mumbling drawl gave expression to a wealth of hard-won ecological knowledge. Wilde had spent years trying to return perennial flow to one of his seasonal streams, even incurring the Forest Service's wrath by cutting down cottonwoods whose roots were sucking up water. Finally he hit upon beaver. Although he live-trapped and released several animals himself over the years, he couldn't persuade them to stick. In 2014 he contacted Wheaton, who agreed to install some Beaver

Dam Analogues to give the next batch of rodents a soft landing. Altogether, nineteen BDAs and five beavers went into Birch Creek in late 2015. The following summer the stream stayed wet two months longer than in past years, sealing Wilde's conversion. "Now I'll put in an earring and grow a ponytail if that's what it takes to get the message out," he told me.

When Wilde, Wheaton, and I drove up a forest road to visit Birch Creek, I immediately understood the rancher's fervor. The stream was a puzzle of pools and ponds, tortuous channels and beaver-dug canals stitching together a soggy castoropolis. Some dams were clearly human-built, others beaver-made; most combined the labor of our two species. Matted grass trails marked routes along which rodents had dragged their aspen victims from forest to pond. Steep valley walls towered above us, creating a funnel that, in spring, would pour rainfall and snowmelt down the defile and into the sponge beneath our feet. Wilde strolled through belly-high grass, caressing seedheads and pronouncing plants' scientific names.

"For a hayseed, you sure know a lot of Latin," I said.

He grinned slyly, bouncing on the balls of his boots. "Can you feel it? The ground?"

I flexed my knees and felt the meadow ripple like Jell-O. "It's soft."

Wilde nodded. "It's saturated."

"I love this so much," Wheaton said. He poked his walking stick into one pond, feeling for the bottom; satisfied it was safe to wade, he slipped in to his chest. "Compared to Bridge Creek, this was child's play," he added. "We just thought about where we wanted to build a primary dam to spread some water onto the floodplain and give the beavers a good lodge location, and then a few secondary dams to extend their forage range." Wheaton's answer to Rosgen's recipe book, I realized, was to help beavers into the kitchen, and then permit them to take over the baking.

We continued to a higher tributary, called Mill Creek, where Beaver Dam Analogues had reached the next phase in their evolution – or devolution. Ordinarily Wheaton and his colleagues constructed BDAs with a hydraulic post pounder, using the jackhammer-like apparatus to drive wooden posts into the creekbed. Although the post pounder made for stable dams, it also precluded building BDAs on remote streams: The device was so heavy that toting its power pack alone required two sets of hands. To avoid the hassle, Wheaton had begun to experiment with "postless BDAs" –

hand-built stick-and-mud mounds that imitated beavers even more faithfully by forgoing machinery altogether.

Wheaton's postless BDAs were blasphemous against stream restoration orthodoxy. The discipline's dominant sect remains focused on stability, installing indomitable rock weirs and bolted log revetments to maintain stream forms over decades. Wheaton's bare-bones beaver dams, by contrast, were ephemeral by design, built to jump-start overbank flows, aggradation, and beaver reestablishment before blowing out during a spring flood. Even then they'd provide prime fish habitat. I recalled Rebekah Levine's words in the Centennial Valley: "One of the coolest things that beaver dams do is fail." Postless BDAs represented the apogee of process-based restoration, in that the dams themselves were almost beside the point. Come back in five years, and they'd likely be gone.

"It's like health care," Wheaton said as we splashed through the pool complex, water tickling my armpits, Wilde watching in amusement from the bank. "The surgeons are the heroes: They sweep in with arrogance and do the reconstructive surgery and walk away. Meanwhile, the patient's in a freakin' cast that the doctor forgets to ever take off. We're trying to do preventive medicine — to use beavers to give rivers a healthy diet of sediment and wood, to let streams exercise a little bit." He turned back and grinned, a bit sheepishly. "Okay, that one is still a work in progress. But you get the idea."

———

Permitting streams to exercise, and employing beavers as their personal trainers, is a radical notion, one that flies in the face of a century of watershed management. Consider the Malheur National Wildlife Refuge. If you've heard of Malheur, you likely know it as the parcel of land in southeast Oregon that, in January 2016, was seized by a slapdash militia of far-right radicals with a decidedly shaky grasp of the Constitution.[9] Nearly a century before Ammon Bundy and his extremist friends laid siege to this high-desert oasis, however, Malheur brimmed with ecological conflict — and beavers were a lead combatant.

Although Peter Skene Ogden and his Hudson's Bay Company trappers ransacked eastern Oregon during their "fur desert" rampage, Malheur's beavers had recovered by the early 1900s, much to the delight of locals. "The beaver, in building dams and developing ponds, increases the supply of pasturage for live stock, creates pond and streams where there is good fishing and recreation in the mountains," enthused William Finley, the biologist who helped win Malheur protection.[10] Such contributions weren't lost on local stockmen. "Mr. Paul Stewart, a rancher near Crane Creek . . . wants very much to have beaver planted upon his place," reported two state employees who spent the summer of 1937 relocating rodents. The transplanters also found a receptive audience in John Scharff, the Malheur refuge's manager, who suggested some relocation sites.[11]

But Scharff's beaver affinity soon hardened into contempt. In the years after World War II, technological triumphalism seized the nation, wildlife biologists included. Producing waterfowl for hunters became the ne plus ultra of wetland management. We'd mastered the atom; surely we could churn out some mallards. With the Department of Agriculture's help, farmers and land managers enlarged and straightened stream channels, tore out riparian plants, and drained some wetlands only to create others — the land management equivalent of better living through chemistry. In 1937 alone, wrote Nancy Langston in her book *Where Land and Water Meet*, Scharff and his crew "built over 159,966 cubic yards of levees and dikes, set 95 miles of barbed wire, cleared out 83,938 cubic yards of channels, laid 34,680 cubic yards of riprap, and set out 35 separate water control structures," channelizing streams and flooding valley bottoms to create bird habitat. "Producing ducks became one more agricultural output," Langston wrote.[12] Anything nature could do, man could better.

The micromanaged version of Malheur had no room for beavers, which burrowed into Scharff's dikes, dammed his channels, and otherwise gummed up the works. Scharff ordered refuge staff to kill twenty-eight beavers in 1946, another forty in 1947, thirty in 1950, and ninety total in 1954 and 1955. Never mind that beavers and waterfowl had cohabitated in harmony for millions of years: Practically overnight, wrote Langston, the rodents became "a hated enemy on the refuge."[13]

While Scharff's specific stream restoration practices may no longer be de rigueur, beavers are still a hated enemy to all those who seek to control nature. Letting streams run wild is an easy prescription on remote rangelands like Jay Wilde's grazing allotments, but it's a considerably more challenging treatment nearer civilization. For every success story like the Walmart beavers of Logan, there are many more cases that resemble the saga of Kelly McAdams and Kris Burns, who, on Christmas Eve 2015, received a hand-delivered citation from the Salt Lake County Division of Flood Control.

McAdams and Burns lived at the time in Draper, a Salt Lake City suburb. Through their backyard flowed Big Willow Creek, a stream that beavers had converted into a rich wetland trafficked by geese, herons, and pelicans. The county was less pleased with the arrangement than the waterfowl were: The creek's beaver dams, claimed the flood control department, presented a threat to downstream property. The citation ordered the couple to remove the structures at once, or face a twenty-five-dollar-a-day fine for as long as the barriers remained standing. "Kind of puts a damper on your Christmas morning," McAdams told me.

Over the next two years, McAdams, Burns, and their neighbors fought the ruling, spending thousands of dollars on appeals and biological surveys. The federal Fish and Wildlife Service even weighed in, requesting in a letter that the Division of Flood Control let the dams stay. But the county would not be dissuaded — the risk was too great, officials insisted. In one *Salt Lake Tribune* story about the esoteric struggle, McAdams vowed to chain himself to the dam, like an Earth Firster locking himself to a spruce.[14] By the time we spoke, though, he and Burns had sold the house and moved to Nevada. "We fought this thing for a couple years, but they were too hardheaded for us," he told me. "We basically capitulated."

It's no surprise that a liability-obsessed flood agency would be loath to accept insubordinate rodents. But the impulse to bend nature to human whims also suffuses more progressive circles: All too often people who live and breathe healthy streams are the same ones lashing out against beavers. Just as John Scharff exterminated Malheur's beavers when they disturbed his duck paradise, domineering approaches to stream restoration still sacrifice the tenacious builders on the altar of progress.

The culprit, in many cases, is a well-intentioned scheme called *compensatory mitigation*. Under the Clean Water Act, industries and agencies, from logging companies to transportation departments, are prohibited from destroying stream and wetland habitat. But there's a loophole: If your construction project promises to cause "unavoidable" damage, you can pay to restore streams elsewhere, like a sixteenth-century Christian buying indulgences to offset sin. The system allows restoration firms to function like house flippers, fixing up degraded streams and selling them off at profit to indulgence-seekers.

As you'd expect, wedging dynamic streams into rigid exchange systems comes with pitfalls. "For the market in stream credits to work, there has to be a defined commodity," Indiana's Rebecca Lave explained to me when I spoke to her for *High Country News* in 2015. Unlike gold or wheat, however, healthy streams are inherently protean; their shifting nature frustrates commodification. "If a stream is changing, regulators have no way to certify whether it's okay or not," Lave said.[15]

The result is that stream mitigation tends to promote stable, simple, salable channels — in short, streams that don't have much room for beavers. Art Parola, the engineer who's reshaping the Winecup Gamble Ranch in Nevada, works primarily in Appalachian states like Kentucky and West Virginia, whose valley floors were once wetted by beaver-filled stream-and-wetland complexes. Beavers were trapped out and their contributions forgotten. These days the Southeast's engineers direct most of their energies toward restoring bottomland hardwood forests — damp stands of oak, hickory, and sycamore surrounding meandering creeks. And heaven help the beaver who interferes with such a habitat.

"If a beaver comes in and makes a pond out of the bottomland hardwood forest you're trying to create, you lose a lot of mitigation credit and your restoration is considered a failure," Parola told me. Not long before we spoke, beavers had converted one of his restored streams into a wetland — a desirable outcome from an ecological standpoint, maybe, but anathema from a regulatory one. He'd been forced to dismantle their lodge, a majestic palace some thirty feet in diameter. "The wood ducks were everywhere," he lamented. "I was almost crying, believe it or not."

Parola's experience speaks to a larger dilemma: Beaver-based restoration is still a new approach, and shoehorning it into existing regulatory systems makes for an uneasy fit. In June 2017 I visited John Coffman, manager of The Nature Conservancy's Red Canyon Ranch in Wyoming, through which sweeps the muddy sine wave of the Little Popo Agie River. Coffman planned to install ten Beaver Dam Analogues on the property later that summer, and had begun to file the raft of paperwork required by the Wyoming Department of Environmental Quality and the Army Corps. Some of the more stringent regulations, he'd noticed, seemed designed to frustrate the very purpose of BDAs. The state had forbidden the structures from exceeding the height or width of the stream under normal flow conditions. "They're worried about the Beaver Dam Analogues diverting water across a meander and creating new channels," laughed Coffman, a bit incredulously, "which is also the goal."

When, many months later, I called Coffman to see how the installations had gone, I wasn't shocked to learn that he hadn't been cleared to build his structures at all. According to the state he now needed to fill out a separate permit requesting the right to use water at each dam — never mind that his BDAs wouldn't actually be withdrawing the water, merely delaying its progress downriver. Coffman still harbored high hopes for the project, but the onerous and head-scratching hang-ups had postponed it for a year. "We got kind of stuck in the mud," he sighed.

For Joe Wheaton and his comrades in process-based restoration, experiences like Coffman's are worrisome harbingers. "Right now we're in this honeymoon period in beaver restoration," Wheaton told me. "Once it gets politicized, the implementation gets so much more difficult. If professional engineers get involved and regulators require these things to be over-designed, that's when it fizzles as a cheap, commonsense solution." Wally Macfarlane, one of Wheaton's Utah State colleagues and a fifth-generation Utahn himself, fears that anti-beaver irrigators and state engineers could weaponize the regulatory process, denying stream alteration permits to dam analogues and shutting down projects before they begin. That's especially concerning in the state's conservative southern half, a region where beavers are about as welcome as university eggheads.

"Many fewer dams exist on the landscape in southern Utah than elsewhere in the state," Macfarlane told me. "There's so much less tolerance. Beavers seem like a slam dunk, but we're getting our shots blocked all the time."

———

The most reliable place to visit beavers in Utah's sunburnt southern country is Mill Creek, the stream that irrigates Moab, recreation capital of the Southwest. When I first emailed Mary O'Brien, a prominent Beaver Believer who lives in nearby Castle Valley, she eagerly agreed to show me the Mill colony's handiwork. Two days before our meeting, though, I opened my inbox to find another communiqué from O'Brien, this one ominously titled

"Uh-oh." As if in confirmation of Wally Macfarlane's warnings, O'Brien had learned that an anonymous vandal had desecrated Mill Creek's dams. Could the damage be repaired? Had the beavers fled the stream? No one knew. Suddenly a serene stroll in the desert had morphed into an episode of *CSI: Moab* — a hunt for a diabolical hooligan and his web-footed victims.

At the appointed dawn, O'Brien and I rambled up Mill Creek, a linear shin-deep oasis that trips through slickrock canyons. The vaulted walls glowed pastel with sunrise: pink, mauve, burnt orange. Willow grew silver along the banks, and cottonwoods waved armfuls of spade-shaped leaves. The sky was white with oncoming heat and we walked in the creek to stay cool, minnows swirling around our sandal-clad feet. At the wet bottom of a dry declivity, we were confronted by the desert's famous contradiction: that a world defined by aridity had been carved by water's passage.

After a few minutes we came to the first dam, or what remained of it. A breach had been torn through the center of the willow wall, ruptured as though with a crowbar, permitting the stream to course through. The channel, released from its impoundment, had cut through layers of reddish silt nearly to bedrock, sweeping truckloads of sediment through the gap. So much material washed down Mill Creek that a popular swimming hole had turned into a mud wallow overnight.

As we continued upstream, every dam we encountered, eight altogether, had been gashed with an all-too-human precision and malice. Later O'Brien heard rumors that the culprit had evicted the beavers for fear, entirely baseless, that they would spread *E. coli*. Although the rodents are so commonly associated with disease that the intestinal malady giardiasis is known as "beaver fever," you're vastly more likely to catch the *Giardia* parasite from a fellow human than from any animal.

Whatever the vandal's motives, he hadn't left any clues. Nor had the beavers. We searched in vain for tracks, fresh cuttings, or signs of repair. "It doesn't look like they're fixing anything up," O'Brien said, hands on hips. "I suspect they're gone from here now." She shook her head in sadness and disgust. "Oh, phooey."

Few people have more experience battling beaver animus than O'Brien, Utah forests program director for a conservation group called the Grand Canyon Trust. O'Brien is a creature of the desert, lean and tough and pragmatic, her deeply tanned face creased by water and wind. She paused often during our hike to deplore invasive plants or bend close to ruddy, well-camouflaged lizards. "You guys are the ones that survive, huh?" she cooed. "The ones that no one can see."

Well adapted though she appears, O'Brien is a relatively recent arrival to Utah, a state where many locals can trace their ancestry to Mormon pioneers. She and her husband moved to Castle Valley in 2003, coaxed from Oregon by their friend, the writer Terry Tempest Williams. "We thought, oh my gosh, here it is — no streetlamps, no businesses, and you can walk out your back door and onto public lands," O'Brien recalled. After a long career in toxics reform, environmental law, and public lands conservation, she wasn't actively looking for work in her new environs, but when the trust came knocking, she came aboard. "In this area, there's a lot of retired people, and they travel and they hike and they hike and they hike," she told me as we, well, hiked. "But too many don't actually *do* anything for the land." That would never, could never, be her.

The trust tasked O'Brien with monitoring Forest Service management plans on the Dixie, Fishlake, and Manti–La Sal National Forests, four and a half million acres of mountains and high plateaus over which cattle ran virtually unrestricted. O'Brien quickly realized the area had been grazed to smithereens. Old aspen, willow, and cottonwood survived on the slopes, but cows and elk devoured new shoots before they could become saplings. Beavers had long since vacated most of the degraded mountain streams. Neat rows of aspen cut at right angles across creeks, like side streets intersecting a main boulevard: a sure indication the trees had taken root atop beaver dams in the distant past. New sign, though, was mostly nonexistent. Beaverland had become a ghost town.

That's not unusual in southern Utah, where cows — more, even, than oil and gas — reign supreme. Three-fifths of Utah's cropland is planted in hay, and livestock consume 70 percent of the water in the Union's second-driest state. Wild animals that threaten cattle's primacy face termination with extreme prejudice. Witness the pitiable coyote, vilified for killing the occasional calf or lamb. The state still pays a fifty-dollar bounty for every pelt hunters turn in, and hosts gruesome coyote-killing derbies that reap hundreds of songdogs.

If there's anything many southern Utahns resent more than coyotes, it's the federal government, which manages around 65 percent of the state's acreage. As in Nevada, state and county officials bridle against what they perceive as bureaucracy's tyranny. The roots of this antipathy are tangled and deep; some run all the way back to the very inception of Utah, whose founders dreamed of ruling an independent Mormon nation. Restrictions on land use are generally met with obloquy. After President Clinton designated Grand Staircase–Escalante National Monument in 1996, state officials launched a smear campaign to discredit the monument – an operation that finally bore fruit in 2017, when Interior Secretary Ryan Zinke resolved to cut Grand Staircase in half.[16] In 2014, after the Bureau of Land Management closed a trail in Recapture Canyon to protect Native American artifacts, a San Juan County commissioner led a motley brigade of off-road vehicles into the wash in protest.[17] "I won't lie – Utah is not an easy place to work as a conservationist," Allison Jones, director of Utah Wild, told me. "You get used to taking baby steps."

Undaunted by the harsh political climate, Mary O'Brien turned the restoration of Utah's beavers into a cause célèbre, throwing festivals in their honor and touting their virtues in *Salt Lake Tribune* op-eds.[18] Her advocacy helped persuade the Utah Division of Wildlife Resources (UDWR) to develop a beaver management plan, a statewide strategy that, among other measures, encourages relocating nuisance rodents instead of killing them. Utah's beaver plan, published in 2010, resolved to improve education efforts, recommended flow devices and other nonlethal techniques, and identified more than a hundred damaged streams as candidates for reloca-tion.[19] The document wasn't perfect – it failed to codify releases on private land, for one thing – but its very existence was remarkable.

As intended, the plan catalyzed a spate of beaver restoration, nowhere more than in the Dixie National Forest, the sweeping aspen plateau that rises above the state's southwest corner. The Dixie is one of the final strong-holds for the region's declining boreal toads – stout, blotchy amphibians that breed, hatch, and metamorphose almost exclusively in beaver ponds. Although historic overgrazing robbed the forest of both aspen and beavers, a round of cattle and elk exclosures, dam analogues, and beaver relocations restored the rodents to the East Fork of the Severe River. Within two years of their return, Mike Golden, a Forest Service biologist in the Dixie, told me, the toads had begun to recolonize ponds where they hadn't ventured in a decade. "That's the amazing thing about beavers when you're a stream ecologist," Golden said. "They do so easily what you have to work so hard to accomplish."

Predictably, not everyone was so enamored of the sudden burst of beaver activity. In 2012 summer arrived in southern Utah so searing and droughty that the US Department of Agriculture added Garfield County to its natural disaster list. Restoring beavers to the Escalante River watershed, which spans Garfield and Kane Counties, seemed to offer a reprieve from drought: The previous year, a consultancy had found that returning the rodents to the Escalante could generate millions of dollars in benefits, water storage among them.[20] Yet the county, in an act of apparent self-sabotage, rejected the state's offer to relocate the wetland-building mammals to its streams. The reason: paranoia that beaver conservation could somehow be exploited to overthrow the bovine hegemony. "We know that it might become a tool for the environmental community to use against cattle," one county chairman told *The Salt Lake Tribune*.[21] Not long after, a Garfield irrigation company killed all thirty-seven beavers living in a single creek in an ill-conceived attempt to hasten the delivery of water.[22]

Although no one mentioned Mary O'Brien's name, no one had to. "I think there were some initial concerns that had more to do with who was involved — some groups rubbed other groups the wrong way, and there was some posturing," Darren DeBloois, the state's furbearer biologist, told me, rather tactfully. When I spoke to DeBloois in September 2017, he said the Department of Wildlife Resources still hoped to negotiate an arrangement with Garfield officials. The county never responded to my requests for comment.

For her part, O'Brien felt like her perspective had been distorted beyond recognition. She didn't consider cattlemen her enemy, she insisted as we drove back to Moab. She'd spent years negotiating collaborative grazing agreements with ranchers and the Forest Service, agreeing to develop new livestock water sources in exchange for keeping cattle out of sensitive creeks. Some stockmen had become close personal friends.

"The Beaver Advisory Committee had representatives from the cattle ranchers and the trappers — they were all cool with it," she said. "But I don't have to be very combative to get a reputation in certain circles. Just the simple idea that beaver should be on the mountains is enough. *Of course* this whole thing is a plan by the Grand Canyon Trust to empty the nation of cows," she groaned, rolling her eyes. "*Of course!*"

"You've got a lot of political forces arrayed against you," I said, stating the obvious.

O'Brien shrugged. "So do climate scientists," she said. "I'm nothing special." By now we'd reached the visitor center in downtown Moab where we'd first rendezvoused, and O'Brien turned the car off and looked at me with a faint smile. "It's like the time I got breast cancer. You know, some people have the reaction 'Why me?' But at the time, I'd been working for eight years in a coalition against pesticides. So my first thought when I was diagnosed, knowing toxics like I do, wasn't 'why me' — it was, 'How did I escape it until now?'" She laughed lightly. "I don't feel sorry for myself. Try living in Syria."

———

The southern Utah saga illustrates an uncomfortable truth at the heart of beaver-based restoration: The landscapes where beavers can do the most good aren't always ready for them. Environmental need and political opportunity don't necessarily overlap, especially in the American West — witness the wolf, at once ecological hero and agricultural villain.

One tool for ensuring better beaver outcomes: BRAT. The Beaver Restoration Assessment Tool is a computer model, developed by Joe Wheaton, Wally Macfarlane, and their Utah State colleagues, that melds publicly available data about stream flow and vegetation to predict how many dams a given creek can support. BRAT is open-source, allowing any rodent-curious researcher to determine where beaver reintroduction or dam analogue construction makes the most sense. Despite its relatively recent invention, BRAT has already become the darling of the castor world. Darren DeBloois, Utah's furbearer biologist, told me the latest iteration of the state's beaver management plan has adopted BRAT to steer its relocation efforts. I've heard beaver advocates in California, Oregon, and Idaho sing BRAT's praises, and Alan Puttock and his Exeter colleagues are developing a version that may someday guide English reintroductions.

Although BRAT's primary value is technical, it is, too, an achievement of the imagination, a method for visualizing the magnificently ponded world that predated European trapping — a time machine to the Castorocene. In 2014 the Utah State team conducted a BRAT appraisal for all of Utah. Once, they found, the state was capable of supporting some 320,000 beaver dams, a whopping 19 per stream mile.[23] Given the ephemerality of beaver complexes, the rodents surely wouldn't have filled every available niche at once. Even at half capacity, though, a dam and pond would have hindered most small streams at every bend.

Since those halcyon days, the tripartite scourge of urbanization, agriculture, and overgrazing has shrunk Utah's dammable potential by around a third. Atop habitat loss, of course, there's trapping. When Wheaton and his colleagues ground-truthed their model, they found the state's streams contained somewhere between 8 and 17 percent as many dams as they could theoretically support — demonstrating that Utah's castorids are operating at a meager fraction of their potential abundance. Some watersheds had nose-dived to just 1 percent.[24] We'll never return to the soggy wonderland that prevailed in 1491, but even a simulacrum would transform the West's waterways for the better. "I think good targets are often in that 30 to 50 percent capacity range," Wheaton told me. "And you can have a hell of a lot of impact just getting it up to 20 or 25 percent."

For that to happen, America's legions of stream restoration professionals may have to undergo a seismic shift — not only in how they perceive beavers, but in how they see themselves. According to a 2017 analysis by Todd BenDor, a professor at the University of North Carolina, America's ecological restoration sector employs 126,000 people and generates nine and a half billion dollars in sales each year[25] — for reference, about twice the revenue of the National Hockey League. Around a third of the Restoration Industrial Complex's workforce labors in streams and wetlands. Beavers present both economic and ideological threats to traditional restoration: They provide highly skilled free labor that obviates the need for expensive interventions, and they serve as four-legged proof that humans do not always know best. Our ecological memories are short-term and faulty; their instincts are unerring and eternal. "If we can hand some of these restoration decisions over to the beaver," Wheaton told me, "we'll be in a much better spot."

———

In 2003 Glenn Albrecht, an Australian philosopher, coined *solastalgia*, a word for the emotional distress we feel at witnessing the destruction of beloved homelands. Albrecht's neologism, which drew from the roots for comfort and pain, had its origins in the territory of New South Wales, where open-pit coal mining destroyed hundreds of square kilometers and inflicted deep psychic wounds on residents. In four syllables *solastalgia* captured the Anthropocene and its discontent: the dissonance of change, the rapidity of loss, the disorientation wrought by environmental grief. "Solastalgia, simply put," Albrecht wrote, "is 'the homesickness you have when you are still at home.'"[26]

One trait that many Beaver Believers share, I've noticed, is a strain of chronic, low-grade solastalgia: homesickness for the ponded, biodiverse world that predated the fur trade. Unlike the solastalgia felt by mining-affected Australians, the homesickness of Beaver Believers is for a home they've never quite known — one they can only re-create through forensic ecology, or snippets of Meriwether Lewis's journals. At risk of melodrama, they remind me sometimes of the American-born children of refugees, dreaming always of a homeland they've never themselves known. What did Beaverland look like, smell like, sound like, feel like beneath bare toes? They're not quite sure, but they miss it anyway.

I grew up in New York's Hudson Valley, a place where the mesh that binds humans to beavers is especially tight. The river, of course, is named after Henry Hudson, one of the fur trade's initiators; between 1630 and 1640 around eighty thousand pelts flowed down the Hudson each year on their way to the Netherlands.[27] New York City's official seal features a Dutch settler, a Lenape Indian, four windmill sails, two barrels of flour, and two beavers. Bennie the Beaver is the mascot of the City College of New York's sports teams. If you disembark from the 6 train at Astor Place and glance up, you'll notice plaques portraying brass beavers in bas-relief — an homage to John Jacob Astor, the real estate magnate and pelt trader whose monopolistic American Fur Company siphoned most of the Rocky Mountains' beavers and eventually helped finance New York's development.

If you wanted to see beavers within city limits, the Astor Place subway stop was long your only option. Although Manhattan Island historically teemed with beavers — Times Square was once a rodent-impounded maple swamp[28] — trapping, development, and pollution pushed them out in the early 1800s. In late 2006, though, a lodge of mud and willow appeared on the banks of the Bronx River. Biologists at the Wildlife Conservation Society named its inhabitant José, after José Serrano, the congressman who'd directed funds toward the river's cleanup. The pioneering rodent became the first wild beaver to grace the Big Apple since before the inception of the New York Police Department.

The person who spends more time on the Bronx River than just about anyone is Katie Lamboy, education coordinator for the Bronx River Alliance, who's led so many boat tours down this beleaguered waterway that fishermen know her as "Canoe Lady." One chill November evening, the Canoe Lady and I pushed off from a rickety dock in the New York Botanical Garden, with her in the bow and me in the stern, to search for José himself. We dipped our paddles into the torpid flow, punting slowly upstream. Beyond the river's wall of trees were Laundromats and chicken joints, pawn shops and bodegas. On the water dusk fell still and quiet, shrouded by red maples and white oaks. Tangles of knotweed screened the banks, and toppled beech trees dipped golden-leaved strainers into the amber river.

José's whereabouts, Lamboy told me, and even whether he survived, were something of a mystery. In 2010 another beaver had shown up, dutifully celebrated by the tabloids — "Lonely Bronx River beaver has a new companion," cheered the *Daily News*.[29] The Wildlife Conservation Society held a naming contest; voters dubbed the new animal Justin Beaver. (Perhaps the only immutable law in beaver restoration is this: If you let children choose names, *Justin Beaver* will always win.) José acquired a Twitter account, @josethebxbeaver, and gained a modest following. Over time, though, people lost track of his movements. "I've actually never seen the beavers," Lamboy sighed. She knew that beavers remained in the Bronx River — a pastor had recently photographed them just upstream of city limits — but not, it seemed, in the Bronx itself.

Spoiler: Katie Lamboy and I did not detect any beavers during our evening on New York City's only freshwater river, though not for a lack of trying. We coasted along the mudslick banks scanning for gnawed stumps or tracks. Nothing. We moored alongside a massive drift of branches deposited on the upstream point of an island, debris jam-packed with PBR cans and candy wrappers and waterlogged textbooks and yellow MetroCards, and examined stick after stick for familiar incisor scars. Nothing. Absence of evidence may not be evidence of absence, but it seemed likely José had split.

Yet in this trash-choked urban channel, there was more than enough nature to satisfy. Lamboy proved a skilled naturalist, adept at spotting red-eared sliders scudding beneath our canoe and circular sunfish nests divoting the sandy bottom. She pointed out a pile of white bivalve shells. "These are invasive Asian clams, but we don't worry about those because the muskrats eat 'em," she said. Around the next bend, sure enough, was the muskrat, her slender tail — a sad thread compared with the beaver's robust appendage — flapping furiously as she slithered into the knotweed. It was easy to forget the river's abused history: the open sewage, the industrial runoff, the limitless trash. I recalled a phrase from Emma Marris's *Rambunctious Garden*: "the sublime in our own backyards."[30]

Katie Lamboy's maternal family moved to the Bronx from Puerto Rico in the 1970s. Soon after they arrived, an uncle drowned while bathing in the river; it became, for Lamboy's relatives and many Bronxites, a place associated with death, pollution, disease. Her grandfather, she told me, "freaked out" when she got her job at the alliance. Mostly, though, her family didn't think about the river at all. "My vision of nature growing up was going to the zoo on Wednesday because it was free," Lamboy told me, her thick black braid swinging as she paddled. "People said to me, you want to study the *environment* here? There *is* no environment in the Bronx."

In 2009 Glenn Albrecht, perhaps ground down by solastalgia's unremitting darkness, concocted an antidote: *soliphilia*, "the notion of political commitment to the saving of loved home environments."[31] I thought of that word when I met Lamboy. Since Howard Cosell declared during the 1977 World Series that the Bronx was burning, the borough, like its river, had been associated with urban decay, a synonym for crack and crime and blight. The decades-long cleanup of the Bronx River became, as José Serrano put it in an NPR interview, "a symbol of the bringing back of life in general to the Bronx" — a way for Bronxites to reclaim the narrative about their own neighborhoods and suffuse it with light.[32] The arrival of José the Beaver made a cute tabloid story, sure, but it was also a new story, a way to convince the world — and residents themselves — that the city had changed for the better. "It said a lot about our resiliency as a borough, and the way we can bounce back, and the pride we have," Lamboy told me as we glided back to the dock. "This beaver is all of us in some way." Coming from anyone else, that might have sounded saccharine; coming from her, it rang wise and genuine and hopeful.

Tell a new friend that you're writing a book about beavers, and you'll often get the same question: "Why, are they endangered?" The answer, we've seen, is no — despite being the targets of a multicentury massacre, millions of hardy rodents flourish today in North American streams, and millions more in Europe. Unlike many wild animals, they endure urbanization, thriving not only in Walmart parking lots, but in stormwater ponds and golf course water hazards. Undeterred by ticking Geiger counters, beavers recolonized the Chernobyl nuclear disaster site within years of the reactor's meltdown. After Mount St. Helens's pyroclastic flows smothered southern Washington beneath volcanic ash, beavers were among the first mammals to reclaim the blast zone — and created habitat for frogs and salamanders among the rubble to boot.[33] As the planet warms and trees encroach on Alaska's once-barren tundra, beavers have followed, happily exploiting climatological catastrophe. (Proving that castorphobia knows no geographic bounds, *The New York Times* labeled the migrants "agents of Arctic destruction" — never mind that beaver-created channels and wetlands may someday help moose, songbirds, and other northward-fleeing species adapt to global warming.[34]) You can have the rats and cockroaches — when the nukes fall, I'm betting on beavers.

At the outset of the twentieth century, beavers were in dire trouble, pushed to within a whisker of obliteration, in desperate need of conservation's aid. Now, in the second decade of the twenty-first, it's we who need *their* help — to store and clean water, to rebuild flood defenses, to repair degraded rivers, to revive biodiversity. Our most powerful environmental law, the Endangered Species Act, works well at protecting rare creatures from disappearing altogether. But there's no parallel requirement that we restore species to their former abundance. Beavers have thoroughly colonized many of our landscapes; reaping the full benefits of their transformative powers will require coexisting with them on millions of acres more. "The conservation of the common," the author J. B. MacKinnon has written, "represents a deeper ambition than the 20th century's lopsided division of the world into islands of wild. . . . It calls on us to integrate conservation into every aspect of human life." [35] Beavers — catholic in their habitat requirements, ark-like in their ability to support other forms of life — represent a spectacular opportunity to practice MacKinnon's ideal. In Lithuania, *Castor fiber* has radically altered the contours of hundreds of agricultural ditches, allowing plants and insects to gain purchase in the "re-naturalized" channels.[36] The only obstacle to returning to the Castorocene is that old hang-up, our cultural carrying capacity — forbearance toward an animal that defies our will.

God, the Old Testament claims, granted man "dominion over the fish of the sea, and over the fowl of the air, and over every living thing that moveth upon the earth," beavers presumably included. Our grandiose sense of self-worth — our human supremacy, as Derrick Jensen has put it — guides our every interaction with other species, from the livestock we raise for slaughter to the wild carnivores whose populations we fanatically control. To work with beavers is to recognize the limits of our own divinely bestowed dominion, to acknowledge that the best thing we can do for many landscapes is to turn their salvation over to a mammal whose ecological vision diverges wildly from our own. *Homo sapiens*'s defining trait is our hubris; ceding our authority to beavers is an act of profound humility. Let the rodent do the work.

Acknowledgments

I n June 2016 I joined a team of National Park Service biologists on a raft-ing trip through the Grand Canyon to study an endangered fish called the razorback sucker. When I emerged twelve days later, so sunburnt that the skin on my calves was sloughing off in sheets, I found awaiting me an email from Michael Metivier, an editor at Chelsea Green Publishing. From my previous writing, Michael had deduced, quite rightly, that I had "a proud love of the beaver," and wondered if I could be convinced to write a book about these world-changing rodents. Accepting his offer proved one of the best decisions I've ever made. Thanks go first, then, to Michael, a smart and joyful steward of *Eager* from conception to comple-tion. This book would not exist without his counsel and good humor. Thanks also to Margo Baldwin and the rest of the Chelsea Green staff for their staunch support.

Becoming a Beaver Believer myself has been an exhilarating intel-lectual journey, one I never would have been able to undertake without skillful guides. All of the experts cited in *Eager* (and several whose work I didn't have space to include) were astonishingly generous with their time and knowledge, and I can't thank them enough for their assistance. Carol Evans, Joe Wheaton, Dan Kotter, Brock Dolman, Kate Lundquist, Mike Callahan, Skip Lisle, Molly Alves, Ruth Shea, and Louise and Paul Ramsay in particular went above and beyond the call of duty in aiding my reporting and travels. Stanley Petrowski and Leonard and Lois Houston of the South Umpqua Rural Community Partnership spearheaded the 2017 State of the Beaver conference in Oregon at which I made many vital contacts. The Houstons have been relocating beavers around southwest Oregon since 2006 and organizing conferences for nearly as long; they're true aquatic rodent champions who are responsible for countless beaver lives saved and human friendships forged.

Two Beaver Believers merit special recognition. First, Kent Woodruff's enthusiasm and generosity have converted dozens to the beaver cult over the years, and I'm proud to be one of his disciples. Second, Heidi Perryman has supplied me an endless stream of stories, sources, studies, and quips since our first email exchange. This book would be far drier without her involvement.

Throughout the writing of *Eager*, the Slackline, my online writers' community, proved a constant source of inspiration, advice, and encouragement. Its many talented members push me daily to work harder, think deeper, and express myself more clearly. Marie Sweetman and Sue Halpern offered financial and logistical input, without which I may never have had the courage to begin this adventure. Keith Hammonds, president of the Solutions Journalism Network, has been an early and consistent supporter of my career, and has taught me much about seeing the whole story. Thanks to everyone who provided me a bed during my reporting, especially longtime friends Terence Lee, Kate Silverman, Monte Kawahara, Cally Carswell, and Sarah Keller. Brian Ertel hosted me in Gardiner, Montana, and coaxed me through a heartstopping grizzly encounter on Specimen Ridge; afterward, Hali Kirby supplied the nerve-calming Moscow Mule. Melody and Daniel Carolan offered Elise and me a snug room in Logan, Utah, and took us hiking among some exuberant wildflowers. Kelsi Sluyter and Eric Rutherford showed me a great time in Pinedale; that I failed to learn how to speed-dance is my fault and mine alone.

Special thanks to the multi-talented Sarah Gilman, who put me up in Portland (again), supplied the gorgeous illustrations for this book, and, once upon a time, edited "The Beaver Whisperer," the *High Country News* seed from which *Eager* ultimately grew. Eric Jay Dolin, the author of the fur trade epic *Fur, Fortune, and Empire*, and the ecologist Bruce Baker perused sections to ensure I wasn't butchering the history and science. Rob Rich, a skilled writer and devout Beaver Believer, provided indispensable edits, both big-picture and granular. I sleep easier knowing the beavers of Swan Valley are in his capable hands.

Much of *Eager* was written at Mi Casita, former New Mexico home of Aldo and Estella Leopold, where I passed a blissful month thanks to a residency from the Aldo Leopold Writing Program. Although Leopold himself wrote little directly about beavers, he would have appreciated

ACKNOWLEDGMENTS

their ability to "preserve the integrity, stability, and beauty of the biotic community"; the opportunity to reread *A Sand County Almanac* on his front porch motivated me to no end. Thanks to program director Richard Rubin, who made me feel right at home in Tres Piedras, and to Leah Todd, a great friend and outstanding journalist who provided incalculable aid during my stay in northern New Mexico. And thanks to another resident of the Land of Enchantment, Dan Flores, who penned *Eager*'s lovely and gracious foreword, and who inspired me with his virtuosic writing on another much-maligned mammal, the coyote.

Thanks, finally, to my family — especially Lisa Saiman, David Goldfarb, Phil Goldfarb, and Shelley Yorke Rose — for their unwavering support, and for tolerating more beaver conversations than any civilians should have to endure. Most of all, thanks to Elise, for her company, wisdom, and love. Since that first fateful trip to the Methow Valley, Elise has journeyed with me to Beaverland and back; I can't imagine writing this book, or doing much of anything, without her. I'm eternally grateful for her companionship, and sustaining her as she's pursued her own dreams has been one of my life's great joys. Here's to the #YearOfTheBeaver, and to many more to come.

Notes

INTRODUCTION

1. *St. Louis Enquirer*, quoted in Eric Jay Dolin, *Fur, Fortune, and Empire* (New York: W. W. Norton, 2010), 224.

2. Osborne Russell, *Journal of a Trapper* (Boise: Syms-York, 1921), 124.

3. Norman MacLean, *A River Runs Through It* (Chicago: University of Chicago Press, 1989), 58.

4. Mark Buckley et al., "The Economic Value of Beaver Ecosystem Services, Escalante River Basin, Utah," case study, ECONorthwest, 2011.

5. J. R. Logan, "Beaver Blamed for Disrupting Taos Cell, Internet Service," *Taos News*, June 27, 2013, http://www.taosnews.com/stories/beaver-blamed-for-disrupting-taos-cell-internet-service,3663.

6. "Beaver Blamed When Tree Strikes Car on Highway," CBC News, July 29, 2014, http://www.cbc.ca/news/canada/prince-edward-island/beaver-blamed-when-tree-strikes-car-on-highway-1.2721282.

7. "Saskatchewan Wedding Left in the Dark After Beaver Bites Down Power Pole," CTV News, June 7, 2017, https://saskatoon.ctvnews.ca/saskatchewan-wedding-left-in-the-dark-after-beaver-bites-down-power-pole-1.3448013.

8. Joseph Goodman, "Birmingham Golf Course Beaver Kill a Dystopian Caddyshack," Alabama Media Group, March 19, 2017, http://www.al.com/sports/index.ssf/2017/03/birmingham_golf_course_beaver.html.

9. Katie Bellis, "Beavers Are No Longer Under Suspicion for Delaying the Filming of the Twin Town Sequel," Wales Online, September 11, 2017, http://www.walesonline.co.uk/news/wales-news/beavers-no-longer-under-suspicion-13604278.

10. Justin Wm. Moyer, "Beaver Walks into Md. Store, Finds Only Artificial Christmas Trees, and Proceeds to Trash It," *Washington Post*, December 1, 2016, https://www.washingtonpost.com/news/local/wp/2016/12/01/cute-beaver-trashes-store-after-holiday-shopping-in-maryland.

11. James B. Trefethen, *An American Crusade for Wildlife* (New York: Winchester Press, 1975), 25.

12. Duncan Halley, Frank Rosell, and Alexander Saveljev, "Population and Distribution of Eurasian Beaver (*Castor fiber*)," *Baltic Forestry* 18, no. 1 (2012): 170.

13. Ernest Thompson Seton, *Lives of Game Animals* (New York: Doubleday, Doran, 1929), 447–48.

14. Ellen Wohl, *Disconnected Rivers* (New Haven, CT: Yale University Press, 2004), 69.

15. Suzanne Fouty, "Climate Change and Beaver Activity," *Beaversprite* 23, no. 1 (2008): 4.

16. Herman Melville, *Moby-Dick; or, The Whale* (New York: Harper & Brothers, 1851), 507.

NOTES

CHAPTER ONE: APPETITE FOR CONSTRUCTION

1. Erwin H. Barbour, "Notice of New Gigantic Fossils," *Science* 19, no. 472 (1892): 99–100.

2. Larry D. Martin, "The Devil's Corkscrew," *Natural History*, April 1994, http://www
.naturalhistorymag.com/htmlsite/editors_pick/1994_04_pick.html.

3. Martin, "The Devil's Corkscrew."

4. Frances Backhouse, *Once They Were Hats* (Toronto: ECW Press, 2015), 38.

5. Michael Runtz, *Dam Builders: The Natural History of Beavers and their Ponds* (Markham,
ON: Fitzhenry & Whiteside, 2015), xxi.

6. Backhouse, *Once They Were Hats*, 18–37.

7. Natalia Rybczynski, "Castorid Phylogenetics: Implications for the Evolution of Swimming
and Tree-Exploitation in Beavers," *Journal of Mammalian Evolution* 14, no. 1 (2007) 1–35.

8. Susanne Horn et al., "Mitochondrial Genomes Reveal Slow Rates of Molecular
Evolution and the Timing of Speciation in Beavers (*Castor*), One of the Largest
Rodent Species," *PLoS One* 6, no. 1 (2011).

9. Anon., "Prosthetic Leg Found in Beaver Dam, Returned to Owner," Associated Press,
August 8, 2016, http://www.carthagepress.com/news/20160808/prosthetic-leg-found
-in-beaver-dam-returned-to-owner.

10. Dietland Müller-Schwarze, *The Beaver: Its Life and Impact*, 2nd ed. (Ithaca, NY:
Cornell University Press, 2011), 55.

11. Lewis Henry Morgan, *The American Beaver and His Works* (Philadelphia: J. B.
Lippincott, 1868).

12. P. I. Danilov and V. Y. Kanshiev, "The State of Populations and Ecological
Characteristics of European (*Castor fiber l.*) and Canadian (*Castor canadensis Kuhl*)
Beavers in the Northwestern USSR," *Acta Zoologica Fennica* 174 (1983): 95–97.

13. Richard R. Buech, David J. Rugg, and Nancy L. Miller, "Temperature in Beaver
Lodges and Bank Dens in a Near-Boreal Environment," *Canadian Journal of Zoology*
67, no. 4 (1989): 1061–66.

14. Kurt M. Samways, Ray G. Poulin, and R. Mark Brigham, "Directional Tree Felling by
Beavers (*Castor canadensis*)," *Northwestern Naturalist* 85 (2004): 48–52.

15. Harold B. Hitchcock, "Felled Tree Kills Beaver (*Castor canadensis*)," *Journal of
Mammalogy* 35, no. 3 (1954): 452.

16. J. L. Harper, "The Role of Predation in Vegetation Diversity," *Brookhaven Symposium
of Biology* 22 (1969): 48–62.

17. Róisín Campbell-Palmer et al., *The Eurasian Beaver Handbook* (Exeter, UK: Pelagic
Publishing, 2015), Kindle edition.

18. Robert J. Gruninger, Tim A. McAllister, and Robert J. Forster, "Bacterial and
Archaeal Diversity in the Gastrointestinal Tract of the North American Beaver
(*Castor canadensis*), *PLoS One* 11, no. 5 (2016).

19. Lyle M. Gordon et al., "Amorphous Intergranular Phases Control the Properties of
Rodent Tooth Enamel," *Science* 347, no. 6223 (2015): 746–50.

20. Ben Johnson, "The Triumph," in *The Oxford Book of English Verse, 1250–1900*, ed.
Arthur Quiller-Couch (Oxford, UK: Oxford University Press, 1900), 178.

21. Charles Wilson, *Notes on the Prior Existence of the Castor fiber in Scotland* (Edinburgh,
Scotland: Neill and Company, 1858), 34.

22. Eric Jay Dolin, *Fur, Fortune, and Empire* (New York: W. W. Norton, 2010), 19.

23. A. Radclyffe Dugmore, *The Romance of the Beaver* (Philadelphia: J. B. Lippincott, 1914), 18.

24. David Coyner, *The Lost Trappers* (Carlise, MA: Applewood Books, 2010), 112.

25. Françoise Patenaude, "Une Année Dans la Vie du Castor," *Les Carnets de Zoologie* 42 (1982): 5–12.

26. Hope Ryden, *Lily Pond: Four Years with a Family of Beavers* (New York: HarperPerennial, 1989).

27. Lars Wilsson, *My Beaver Colony* (New York: Doubleday, 1968).

28. Dugmore, *Romance of the Beaver*, 45.

29. Morgan, *American Beaver and His Works*, 18.

30. Morgan, *American Beaver and His Works*, 106.

31. F. G. Speck, "The Basis of Indian Ownership of Land and Game," in *The Southern Workman* (Hampton, VA: Press of the Hampton Normal and Agricultural Institute, 1914), 36.

32. David Malakoff, "150-Year-Old Map Reveals That Beaver Dams Can Last Centuries," *Science*, December 4, 2015, http://www.sciencemag.org/news/2015/12/150-year-old-map-reveals-beaver-dams-can-last-centuries.

33. Carol A. Johnston, "Fate of 150-Year-Old Beaver Ponds in the Laurentian Great Lakes Region," *Wetlands* 35, no. 5 (2015): 1018.

34. John McPhee, "Atchafalaya," *The New Yorker*, February 23, 1987, https://www.newyorker.com/magazine/1987/02/23/atchafalaya.

35. Lina E. Polvi and Ellen Wohl, "The Beaver Meadow Complex Revisited – The Role of Beavers in Post-Glacial Floodplain Development," *Earth Surface Processes and Landforms* 37, no. 3 (2012): 332–46.

36. M. Gordon Wolman and Luna B. Leopold, "River Flood Plains: Some Observations on Their Formation," *Physiographic and Hydraulic Studies of Rivers*, Geological Survey Professional Paper 282-C (1957).

37. Suzanne C. Fouty, "Euro-American Beaver Trapping and Its Long-Term Impact on Drainage Network Form and Function, Water Abundance, Delivery, and System Stability," in *Riparian Research and Management: Past, Present, Future*, General Technical Report, ed. R. Roy Johnson, Steven W. Carothers, Deborah M. Finch, Kenneth J. Kingsley, and John T. Stanley (Fort Collins: US Forest Service, Rocky Mountain Research Station, 2018), 1:102–33, https://doi.org/10.2737/RMRS-GTR-377-CHAP7.

38. Denise Burchsted et al., "The River Discontinuum: Applying Beaver Modifications to Baseline Conditions for Restoration of Forested Headwaters," *BioScience* 60, no. 11 (2010): 908–22.

39. David R. Butler and George P. Malanson, "The Geomorphic Influence of Beaver Dams and Failures of Beaver Dams," *Geomorphology* 71, nos. 3–4 (2005): 56.

CHAPTER TWO: DISLODGED

1. John Kirk Townsend, *Narrative of a Journey Across the Rocky Mountains* (Philadelphia: Henry Perkins, 1839), 75.

2. John Kirk Townsend, *Sporting Excursions in the Rocky Mountains*, vol. 1 (London: Henry Colburn, 1840), 131.

3. Osborne Russell, *Journal of a Trapper* (Boise: Syms-York, 1921), 142.

4. Washington Irving, *Memoirs of Washington Irving: With Selections from His Works, and Criticisms* (New York: Carlton & Lanahan, 1870), 237.

5. David Coyner, *The Lost Trappers* (1847; repr., Carlise, MA: Applewood Books, 2010), 219.

6. Oscar Wilde, *The Ballad of Reading Gaol and Other Poems* (New York: Dover Publication), 26.

7. Shannon Hengen and Ashley Thompson, *Margaret Atwood: A Reference Guide* (Lanham, MD: Scarecrow Press, 2007), 205.

8. James Truslow Adams, *The Founding of New England* (Boston: Atlantic Monthly Press, 1921), 102.

9. Eric Jay Dolin, *Fur, Fortune, and Empire* (New York: W. W. Norton, 2010), 121.

10. Charlotte Erichsen-Brown, *Medicinal and Other Uses of North American Plants* (New York: Dover Publications, 1989), xi.

11. Robin S. Doak, *Subarctic Peoples* (Chicago: Heinemann Library, 2012), 25.

12. Harold Hickerson, "Fur Trade Colonialism and the North American Indian," *Journal of Ethnic Studies* 1 (Summer 1973), 39.

13. Dolin, *Fur, Fortune, and Empire*, 34.

14. Louis Rodrigue Masson, *Les Bourgeois de la Compagnie du Nord-Ouest* (Quebec: De L'Imprimerie Generale, 1880), 342.

15. Harold Innis, *The Fur Trade in Canada* (1930; repr., New Haven, CT: Yale University Press, 1962), 187.

16. Captain Marryat, *A Diary in America with Remarks and Illustrations, Part Second,* vol. 3 (London: Longman, Orme, Brown, Green, and Longmans, 1839), 228.

17. Donald Berry, *A Majority of Scoundrels* (New York: Harper and Brothers, 1961), 18.

18. Mari Sandoz, *The Beaver Men* (Lincoln: University of Nebraska Press, 1964), 86.

19. John Bakeless, *The Eyes of Discovery* (Philadelphia: J. B. Lippincott Company, 1950), 245.

20. Bakeless, *The Eyes of Discovery*, 245.

21. Daniel Wilson, "Early Notices of the Beaver, in Europe and North America," in *The Canadian Journal of Industry, Science, and Art*, vol. 4 (Toronto: Printed for the Canadian Institute by Lovell and Gibson, 1859), 363.

22. Harry V. Radford, *Artificial Preservation of Timber and History of the Adirondack Beaver* (Albany: J. B. Lyon, 1908), 396.

23. Bakeless, *The Eyes of Discovery*, 245.

24. Bakeless, *The Eyes of Discovery*, 293.

25. Pierre Esprit Radisson and Gideon Scull, *Voyages of Pierre Esprit Radisson* (Boston: Prince Society, 1885), 191–92.

26. Meriwether Lewis, *Journals of the Lewis and Clark Expedition* (University of Nebraska Press, https://lewisandclarkjournals.unl.edu), June 30, 1805.

27. Charles G. Clarke, *The Men of the Lewis and Clarke Expedition* (Lincoln: University of Nebraska Press, 1970), 183.

28. Lewis, *Journals of the Lewis and Clark Expedition*, July 18, 1805.

29. Lewis, *Journals of the Lewis and Clark Expedition*, July 24, 1805.

30. Lewis, *Journals of the Lewis and Clark Expedition*, July 30, 1805.

31. Lewis, *Journals of the Lewis and Clark Expedition*, August 2, 1805.

32. Arlen J. Large, "Expedition Aftermath: The Jawbone Journals," *We Proceeded On* 17, no. 1 (February 1991): 13–25.

33. Dolin, *Fur, Fortune, and Empire*, 180.

34. Rosalyn LaPier, "The Piegan View of the Natural World" (PhD diss., University of Montana, 2015), 200.

35. R. Grace Morgan, "Beaver Ecology/Beaver Mythology" (PhD diss., University of Alberta, 1991), 61.

36. Morgan, "Beaver Ecology/Beaver Mythology," 6.

37. John James Audubon, *The Missouri Journals* (Wikisource: https://en.wikisource.org/wiki/Audubon_and_His_Journals/The_Missouri_River_Journals), May 12, 1843.

38. Charles Pierce, "JFK at 86," *Esquire Magazine*, January 29, 2007, http://www.esquire.com/news-politics/news/a457/esq1003-oct-jfk.

39. Henry David Thoreau, *Thoreau's Animals*, ed. Geoff Wisner (New Haven, CT: Yale University Press, 2017), 23.

40. William Cronon, *Changes in the Land* (New York: Hill and Wang, 1983), 106.

41. James Campbell Lewis and George Edward Lewis, *Black Beaver the Trapper* (George Edward Lewis, 1911), 28.

42. Jeremy Belknap, *The History of New Hampshire*, vol. 3 (Boston: Bradford and Read, 1813), 118–19.

43. Henry Wansey, *An Excursion to the United States of America in the Summer of 1794* (Salisbury, UK: J. Easton, 1798), 197.

44. Suzanne C. Fouty, "Euro-American Beaver Trapping and Its Long-Term Impact on Drainage Network Form and Function, Water Abundance, Delivery, and System Stability," in *Riparian Research and Management: Past, Present, Future* (US Forest Service, 2018), 1:102–33.

45. Shirley T. Wajda, "Ending the Danbury Shakes: A Story of Workers' Rights and Corporate Responsibility," connecticuthistory.org.

46. Daniel P. Jones, "Mad Hatter's Legacy," *Hartford Courant*, September 22, 2002, http://articles.courant.com/2002-09-22/news/0209221170_1_mercury-levels-factory-workers-disease-registry.

47. Johan Varekamp and Daphne Varekamp, "Adriaen Block, the Discovery of Long Island Sound and the New Netherlands Colony: What Drove the Course of History?" *Wrack Lines Magazine*, Connecticut Sea Grant, Summer 2006.

48. Johan Varekamp et al., "Environmental Change in the Long Island Sound in the Recent Past: Eutrophication and Climate Change," Long Island Sound Research Fund, Connecticut Department of Environmental Protection, January 20, 2010.

49. Carol Kuhn, "Naturalists in the Rocky Mountain Fur Trade: 'They Are a Perfect Nuisance,'" *Rocky Mountain Fur Trade Journal* 10 (2016): 78.

50. John Moring, *Early American Naturalists: Exploring the American West, 1804–1900* (Lanham, MD: Taylor Trade Publishing, 2005), 55.

51. John Kirk Townsend, quoted in Kuhn, "Naturalists in the Rocky Mountain Fur Trade," 77.

52. Rudolf Ruedemann and W. J. Schoonmaker, "Beaver Dams as Geologic Agents," *Science* 88, no. 2292 (1938): 523–25.

53. Lina E. Polvi and Ellen Wohl, "The Beaver Meadow Complex Revisited – The Role of Beavers in Post-Glacial Floodplain Development," *Earth Surface Processes and Landforms* 37, no. 3 (2012): 332–46.

54. Backhouse, *Once They Were Hats*, 3.

55. Hilary A. Cooke and Steve Zack, "Influence of Beaver Dam Density on Riparian Areas and Riparian Birds in Shrubsteppe of Wyoming," *Western North American Naturalist* 68, no. 3 (2008): 370.

56. Robert J. Naiman, Carol A. Johnston, and James C. Kelley, "Alteration of North American Streams by Beaver," *BioScience* 38, no. 11 (1988): 753–62.

57. Robert J. Naiman et al., "Beaver Influences on the Longterm Biogeochemical Characteristics of Boreal Forest Drainage Networks," *Ecology* 75, no. 4 (1994): 907.

58. Naiman, Pinay, Johnston, and Pastor, "Beaver Influences," 918.

59. William Temple Hornaday, quoted in Tim Lehman, *Bloodshed at Little Bighorn: Sitting Bull, Custer, and the Destinies of Nations* (Baltimore: Johns Hopkins University Press, 2010), 59.

60. Frank Graham Jr., *Man's Dominion: The Story of Conservation in America* (New York: M. Evans, 1971), 14.

61. Cronon, *Changes in the Land*, 120.

62. Cronon, *Changes in the Land*, 121.

63. Rod Giblett, *Black Swan Lake: Life of a Wetland* (Bristol, UK: Intellect, 2013), 187.

64. Roberta H. Yuhas, "Loss of Wetlands in the Southwestern United States," abstracted from US Geological Survey Water-Supply Paper 2425, National Water Summary on Wetland Resources, 1996, https://geochange.er.usgs.gov/sw/impacts/hydrology/wetlands.

65. Aldo Leopold, *A Sand County Almanac, and Sketches Here and There* (New York: Oxford University Press, 1949), 162.

66. Robert L. France, "The Importance of Beaver Lodges in Structuring Littoral Communities in Boreal Headwater Lakes," *Canadian Journal of Zoology* 75, no. 7 (1997): 1009–13.

67. Duffy J. Brown, Wayne A. Hubert, and Stanley H. Anderson, "Beaver Ponds Create Wetland Habitat for Birds in Mountains of Southeastern Wyoming," *Wetlands* 16, no. 2 (1996): 127–33.

68. Brian S. Metts, J. Drew Lanham, Kevin R. Russell, "Evaluation of Herpetofaunal Communities on Upland Streams and Beaver-Impounded Streams in the Upper Piedmont of South Carolina," *American Midland Naturalist* 145, no. 1 (2001): 54–65.

69. Justin P. Wright, Clive G. Jones, and Alexander S. Flecker, "An Ecosystem Engineer, the Beaver, Increases Species Richness at the Landscape Scale," *Oecologia* 132, no. 1 (2002): 96–101.

70. Joseph M. Smith and Martha E. Mather, "Beaver Dams Maintain Fish Biodiversity by Increasing Habitat Heterogeneity Throughout a Low-Gradient Stream Network," *Freshwater Biology* 58, no. 7 (2013): 1523–38.

71. George J. Knudsen, "Relationship of Beaver to Forests, Trout, and Wildlife in Wisconsin," Technical Bulletin No. 25, Wisconsin Conservation Department (1962).

72. Victoria H. Zero and Melanie A. Murphy, "An Amphibian Species of Concern Prefers Breeding in Active Beaver Ponds," *Ecosphere* 7, no. 5 (2016).

73. Michael M. Pollock, Robert J. Naiman, and Thomas A. Hanley, "Plant Species Richness in Riparian Wetlands – A Test of Biodiversity Theory," *Ecology* 79, no. 1 (1998): 94–105.

74. Nis L. Anderson, Cynthia A. Paszkowski, and Glynnis A. Hood, "Linking Aquatic and Terrestrial Environments: Can Beaver Canals Serve as Movement Corridors for Pond-Breeding Amphibians?" *Animal Conservation* 18, no. 3 (2008): 287–94.

75. Joseph K. Bailey and Thomas G. Whitham, "Interactions Between Cottonwood and Beavers Positively Affects Sawfly Abundance," *Ecological Entomology* 31, no. 4 (2006): 294–97.

76. William Byrd, *The Westover Manuscripts: Containing the History of the Dividing Line Betwixt Virginia and North Carolina* (Petersburg, VA: Edmund and Julian C. Ruffin, 1841), 44.

77. James A. Hanson, *When Skins Were Money: A History of the Fur Trade* (Chadron, NE: Museum of the Fur Trade, 2005), 49.

78. Ann Vileisis, *Discovering the Unknown Landscape: A History of America's Wetlands* (Washington, DC: Island Press, 1997), 48.

79. Lora Griffiths, "Weekend at a Glance: Beaver Queen, Bulls and Blues," *Durham Magazine*, June 1, 2017, https://durhammag.com/2017/06/01/weekend-at-a-glance-june-1-4.

80. Heather Cayton et al., "Habitat Restoration as a Recovery Tool for a Disturbance-Dependent Butterfly, the Endangered St. Francis' Satyr," *Butterfly Conservation in North America*, ed. Jaret C. Daniels (Dordrecht, Netherlands: Springer, 2015): 157.

81. Justin Gillis, "Let Forest Fires Burn? What the Black-Backed Woodpecker Knows," *New York Times*, August 6, 2017, https://www.nytimes.com/2017/08/06/science/let-forest-fires-burn-what-the-black-backed-woodpecker-knows.html?_r=0.

82. James B. Trefethen, *An American Crusade for Wildlife* (New York: Winchester Press, 1975), 25.

83. Mark E. Harmon, "Moving Toward a New Paradigm for Woody Detritus Management," *Ecological Bulletins* 49 (2001): 269–278.

84. Stella Thompson, Mia Vehkaoja, and Petri Nummi, "Beaver-Created Deadwood Dynamics in the Boreal Forest," *Forest Ecology and Management* 360 (2016): 1–8.

85. Andrew C. Revkin, "An Anthropocene Journey," *Anthropocene Magazine*, October 2016, http://www.anthropocenemagazine.org/anthropocenejourney.

86. Joseph K. Bailey et al., "Beavers as Molecular Geneticists: A Genetic Basis to the Foraging of an Ecosystem Engineer," *Ecology* 85, no. 3 (2004): 603–8.

CHAPTER THREE: DECEIVE AND EXCLUDE

1. Robert Michael Ballantyne, *Hudson's Bay: Or Everyday Life in the Wilds of North America* (Boston: Philips, Sampson, 1859), 42.

2. George Bird Grinnell, "Spare the Trees," quoted in John Reiger, "Pathbreaking Conservationist," *Forest History Today*, Spring–Fall 2005, 19.

3. Gifford Pinchot, quoted in Jacqueline Vaughn Switzer, *Green Backlash: The History and Politics of Environmental Opposition in the U.S.* (Boulder, CO: Lynne Rienner Publishers, 1997), 191.

4. Enos Mills, *In Beaver World* (Boston and New York: Houghton Mifflin, 1913), 29.

5. Mills, *In Beaver World*, 29–30.

6. Mills, *In Beaver World*, 213.

7. Arthur Chapman, "Enos Mills: Nature Guide," excerpted online from *Enos A. Mills, Author, Speaker, Nature Guide* (Longs Peak, CO: Trail Book Store, 1921), http://www.enosmills.com/exasng.pdf.

8. Byron Anderson, "Biographical Portrait: Enos Abijah Mills," *Forest History Today*, Spring–Fall 2007, 57.

9. Enos Mills, *The Adventures of a Nature Guide* (Garden City, NY: Doubleday, Page, 1920), 174.

10. Mills, *In Beaver World*, 221.

11. John Warren, "Extinction: A Short History of Adirondack Beaver," *Adirondack Almanack*, April 8, 2009, https://www.adirondackalmanack.com/2009/04/extinction-a-short-history-of-adirondack-beaver.html.

12. Warren, "Extinction."

13. John F. Organ et al., "A Case Study in the Sustained Use of Wildlife: The Management of Beaver in the Northeastern United States," in *Enhancing Sustainability: Resources for Our Future*, ed. Hendrik A. van der Linde and Melissa H. Danskin (Cambridge, UK: International Union for Conservation of Nature and Natural Resources, 1998), 125–40.

14. J. P. Walker, *The Legendary Mountain Men of North America* (lulu.com, 2015), 176, https://books.google.com/books/about/The_Legendary_Mountain_Men_of_North_Amer.html?id=SUlpCQAAQBAJ.

15. Kate Lundquist and Brock Dolman, "Beaver in California: Creating a Culture of Stewardship," Occidental Arts and Ecology Center WATER Institute, 2016, 7.

16. Organ, Gotie, Decker, and Batcheller, "A Case Study in the Sustained Use of Wildlife," 126.

17. Stanley E. Hedeen, "Brief Note: Return of the Beaver, *Castor canadensis*, to the Cincinnati Region," *Ohio Journal of Science* 85, no. 4 (1985): 202–3.

18. Ruth M. Elsey and Noel Kinler, "Range Extension of the American Beaver, *Castor canadensis*, to Louisiana," *The Southwestern Naturalist* 41, no. 1 (1996): 91–93.

19. Organ, Gotie, Decker, and Batcheller, "A Case Study in the Sustained Use of Wildlife," 126.

20. Jim Sterba, *Nature Wars* (New York: Crown Publishers, 2012), 75.

21. Allan E. Houston, "The Beaver — A Southern Native Returning Home," in *Proceedings of the 18th Vertebrate Pest Conference* (University of California–Davis, 1998), 16.

22. Angie Bevington, "Frank Ralph Conibear," *Arctic Profiles* (1983): 386–87. http://pubs.aina.ucalgary.ca/arctic/Arctic36-4-386.pdf.

23. "Human-Beaver Conflicts in Massachusetts: Assessing the Debate over Question One," Massachusetts Society for the Prevention of Cruelty to Animals, The Humane Society of the United States, Animal Protection Institute and United Animal Nations, 2005, http://www.humanesociety.org/assets/pdfs/MA_beaver_report_2005.pdf.

24. Sterba, *Nature Wars*, 59–85.

25. "Human-Beaver Conflicts in Massachusetts."

26. Paul Frisman, "Massachusetts' Ban on Leg-Hold Traps," *OLR Research Report*, January 23, 2001, https://www.cga.ct.gov/2001/rpt/2001-R-0127.htm.

27. "Human-Beaver Conflicts in Massachusetts."

28. Laura J. Simon, "Solving Beaver Flooding Problems Through the Use of Water Flow Control Devices," in *Proceedings of the 22nd Vertebrate Pest Conference* (University of California–Davis, 2006), 174–80.

29. Bob Salsberg, "Survey Finds Most New England Culverts Aren't Meeting Standards," Associated Press, August 6, 2011, http://bangordailynews.com/2011/08/06/environment/survey-finds-most-new-england-culverts-are-subpar.

30. "Human-Beaver Conflicts in Massachusetts."

31. Cordelia Zars, "With Pelt Prices Dropping, N.H.'s Beaver Population Grows," New Hampshire Public Radio, August 17, 2016, http://nhpr.org/post/pelt-prices-dropping-nhs-beaver-population-grows.

32. Animal and Plant Health Inspection Service, US Department of Agriculture, "Program Data Report G: Animals Dispersed / Killed or Euthanized / Removed or Destroyed / Freed," 2016, https://www.aphis.usda.gov/wildlife_damage/pdr/PDR-G_Report.php.

33. "Decision and Finding of No Significant Impact: Environmental Assessment — Reducing Aquatic Rodent Damage Through an Integrated Wildlife Damage Management Program in the Commonwealth of Virginia," 2007. Unpublished. Accessed through Freedom of Information Act request.

34. Dale L. Nolte, Seth R. Swafford, and Charles A. Sloan, "Survey of Factors Affecting the Success of Clemson Beaver Pond Levelers Installed in Mississippi by Wildlife Services," in *The Ninth Wildlife Damage Management Conference Proceedings* (2000), 120–25.

35. Stephanie L. Boyles and Barbara A. Savitzky, "An Analysis of the Efficacy and Comparative Costs of Using Flow Devices to Resolve Conflicts with North American Beavers Along Roadways in the Coastal Plain of Virginia," in *Proceedings of the 23rd Vertebrate Pest Conference* (University of California–Davis, 2008), 47–52.

36. Glynnis A. Hood, Varghese Manaloor, and Brendan Dzioba, "Mitigating Infrastructure Loss from Beaver Flooding: A Cost-Benefit Analysis," *Human Dimensions of Wildlife* 23, no. 2 (2018): 156.

37. Mike Callahan, "Best Management Practices for Beaver Problems," *Association of Massachusetts Wetland Scientists Newsletter* 53 (2005): 13.

38. Mike Callahan, "Beaver Management Study," *Association of Massachusetts Wetland Scientists Newsletter* 44 (2003): 15.

39. Bart Baca et al., "Economic Analyses of Wetlands Mitigation Projects in the Southeastern U.S.," in *Proceedings of the 21st Annual Conference on Wetlands Restoration and Creation* (1994), 16–24.

40. Hood, Manaloor, and Dzioba, "Mitigating Infrastructure Loss from Beaver Flooding," 10.

41. In fact, the weight of a beaver's brain is 41 to 45 grams, about the weight of a golf ball. Still, point taken. See Dietland Müller-Schwarze and Lixing Sun, *The Beaver: Natural History of a Wetlands Engineer* (Ithaca, NY: Cornell University Press, 2003), 11.

CHAPTER FOUR: THE BEAVER WHISPERER

1. "Beaver Management," Substitute House Bill 2349, 62nd Legislature, Washington State, filed March 29, 2012.

2. Elmo W. Heter, "Transplanting Beavers by Airplane and Parachute," *Journal of Wildlife Management* 14, no. 2 (1950): 144.

3. Elmo W. Heter, "Transplanting Beavers by Airplane and Parachute," 146.

4. Elmo W. Heter, "Transplanting Beavers by Airplane and Parachute," 146.

5. Mark C. McKinstry and Stanley H. Anderson, "Attitudes of Private- and Public-Land Managers in Wyoming, USA, Toward Beaver," *Environmental Management* 23, no. 1 (1999): 98.

6. Mark McKinstry and Stanley H. Anderson, "Survival, Fates, and Success of Transplanted Beavers, *Castor canadensis*, in Wyoming," *Canadian Field Naturalist* 116, no. 1 (2002): 60–68.

7. Mark McKinstry, Paul Caffrey, and Stanley H. Anderson, "The Importance of Beaver to Wetland Habitats and Waterfowl in Wyoming," *Journal of the American Water Resources Association* 37, no. 6 (2001): 1573.

8. McKinstry, Caffrey, and Anderson, "The Importance of Beaver to Wetland Habitats and Waterfowl in Wyoming," 1573.

9. Marina Milojevic, "Castoreum," *Fragrantica*, https://www.fragrantica.com/notes/Castoreum-102.html.

10. Baron Ambrosia, "Tales From the Fringe: Beaver Gland Vodka," *Punch*, February 6, 2015, https://punchdrink.com/articles/tales-from-the-fringe-beaver-gland-vodka/.

11. Lixing Sun and Dietland Müller-Schwarze, "Sibling Recognition in the Beaver: A Field Test for Phenotype Matching," *Animal Behavior* 54, no. 3 (1997): 493–502.

12. Michael M. Pollock et al., *The Beaver Restoration Guidebook: Working with Beaver to Restore Streams, Wetlands, and Floodplains*, version 1.0, US Fish and Wildlife Service (Portland, OR, 2015), 74.

13. Dietland Müller-Schwarze, *The Beaver: Its Life and Impact*, 2nd ed. (Ithaca, NY: Cornell University Press, 2011), 45.

14. Vanessa Petro, "Evaluating 'Nuisance' Beaver Relocation as a Tool to Increase Coho Salmon Habitat in the Alsea Basin of the Central Oregon Coast Range" (master's thesis, Oregon State University, 2013).

15. David S. Pilliod et al., "Survey of Beaver-Related Restoration Practices in Rangeland Streams of the Western USA," *Environmental Management* 61, no. 1 (2018): 58–68.

16. Jeremiah Wood, "2017 Fur Prices: NAFA July Auction Results," *Trapping Today*, July 12, 2017, http://trappingtoday.com/2017-fur-prices-nafa-july-auction-results.

17. Martin Nie et al., "Fish and Wildlife Management on Federal Lands: Debunking State Supremacy," *Environmental Law* 47, no. 4 (2017), https://ssrn.com/abstract=2980807.

18. Anon., "Chapter Four: Furbearing Animal Hunting or Trapping Seasons," Wyoming Department of Game and Fish, November 9, 2016, https://wgfd.wyo.gov/Regulations/Regulation-PDFs/REGULATIONS_CH4.pdf.

19. Diana Hembree, "Cattle Ranchers Join Conservationists to Save Endangered Species and Rangelands," *Forbes*, January 5, 2018, https://www.forbes.com/sites/dianahembree/2018/01/05/cattle-ranchers-join-conservationists-to-save-endangered-species-rangelands/#4e225dbe220d.

20. "Idaho State Wildlife Action Plan, 2015," Idaho Department of Fish and Game, Sponsored by the US Fish and Wildlife Service, completed under grant F14AF01068 Amendment #1, 2017.

21. Anon., "Game Species, Small Game and Trapping, Beaver Management," Washington Department of Fish and Wildlife, updated January 2018. https://wdfw.wa.gov/hunting/trapping/beaver/.

22. Nathan Halverson, "9 Sobering Facts About California's Groundwater Problem," *Reveal News*, June 25, 2015, https://www.revealnews.org/article/9-sobering-facts-about-californias-groundwater-problem.

23. Daniel J. Karran, Cherie J. Westbrook, and Angela Bedard-Haughn, "Beaver-Mediated Water Table Dynamics in a Rocky Mountain Fen," *Ecohydrology* 11, no. 2 (2018).

24. Cherie J. Westbrook, David J. Cooper, and Bruce W. Baker, "Beaver Dams and Overbank Floods Influence Groundwater-Surface Water Interactions of a Rocky Mountain Riparian Area," *Water Resources Research* 42, no. 6 (2006).

25. The Lands Council, "The Beaver Solution: An Innovative Solution for Water Storage and Increased Late Summer Flows in the Columbia River Basin," completed under Washington Department of Ecology Grant #G0900156.

26. Keith Ridler, "Hot Water Kills Half of Columbia River Sockeye Salmon," Associated Press, July 27, 2015, http://www.oregonlive.com/environment/index.ssf/2015/07/hot_water_killing_half_of_colu.html.

27. Ralph Maughan, "Beaver Restoration Would Reduce Wildfire," *Wildlife News*, October 25, 2013, http://www.thewildlifenews.com/2013/10/25/beaver-restoration-would-reduce-wildfires.

CHAPTER FIVE: REALM OF THE DAMMED

1. Timothy Egan, *The Good Rain* (New York: Vintage Books, 2011), 182.

2. Bernie Gobin Kia-Kia, "The Story of the Salmon Ceremony," Stories and Teachings, the Tulalip Tribes. Accessed February 6, 2018. https://tulaliptoday.com/about/stories-teachings/crane-and-changer/.

3. James Wickersham, quoted in Ezra Meeker, *Pioneer Reminiscences of Puget Sound* (Seattle, WA: Lowman & Hanford, 1905), 251.

4. Paul Cereghino et al., "Strategies for Nearshore Protection and Restoration in Puget Sound," prepared in support of the Puget Sound Nearshore Ecosystem Restoration Project, March 2012.

5. David Montgomery, *King of Fish: The Thousand-Year Run of Salmon* (Cambridge, MA: Westview Press, 2004), 208.

6. Robert J. Naiman, Carol A. Johnston, and James C. Kelley, "Alteration of North American Streams by Beaver," *BioScience* 38, no. 11 (1988): 753–62.

7. Glennda Chui, "Jammin' for the Salmon," *San Jose Mercury News*, January 26, 1999, https://www.wcc.nrcs.usda.gov/ftpref/wntsc/strmRest/EngineeredLogJamsForSalmon.pdf.

8. Montgomery, *King of Fish*, 217.

9. Holly Labadie, "Beaver Dam Management Project — 2015," Miramichi Salmon Association, November 24, 2015.

10. Tom Knudson, "Native Beavers Suffer in Tahoe as USFS Protects Non-Native Kokanee," *Lake Tahoe News* (originally published in *Sacramento Bee*), October 7, 2012, http://www.laketahoenews.net/2012/10/native-beavers-suffer-in-tahoe-as-usfs-protect-non-native-kokanee.

11. Daryl Guignion, "A Conservation Strategy for Atlantic Salmon in Prince Edward Island," Prince Edward Island Council of the Atlantic Salmon Federation, March 2009.

12. Sharon T. Brown, "Atlantic Salmon/Beaver Dam Controversy," *Beaversprite*, Summer 2009, http://www.beaversww.org/assets/PDFs/AtlanticSalmonBeaverDam.pdf.

13. Anon., "Wisconsin Beaver Management Plan, 2015–2025," Wisconsin Department of Natural Resources, 2015, 53.

14. Ed L. Avery, "Fish Community and Habitat Responses in a Northern Wisconsin Brook Trout Stream 18 Years After Beaver Dam Removal," Wisconsin Department of Natural Resources, Bureau of Integratd Science Services, April 1, 2002.

15. Sharon T. Brown, "Wisconsin's War on Nature," *Beaversprite*, Winter 2011–12, http://www.beaversww.org/assets/PDFs/Wisconsins-War-on-Nature-1.pdf.

16. Gil McRae and Clayton J. Edwards, "Thermal Characteristics of Wisconsin Headwater Streams Occupied by Beaver: Implications for Brook Trout Habitat," *Transactions of the American Fisheries Society* 123, no. 4 (1994): 641–56.

17. Ryan L. Lokteff, Brett B. Roper, and Joe M. Wheaton, "Do Beaver Dams Impede the Movement of Trout?" *Transactions of the American Fisheries Society* 142, no. 4 (2013): 1114–25.

18. Seth M. White and Frank J. Rahel, "Complementation of Habitats for Bonneville Cutthroat Trout in Watersheds Influenced by Beavers, Livestock, and Drought," *Transactions of the American Fisheries Society* 137, no. 3 (2008): 881–94.

19. Naiman, Johnston, and Kelley, "Alteration of North American Streams by Beaver."

20. D. Gorshkov, "Is It Possible to Use Beaver Building Activity to Reduce Lake Sedimentation?" *Lutra* 46, no. 2 (2003): 189–96.

21. W. Gregory Hood, "Beaver in Tidal Marshes: Dam Effects on Low-Tide Channel Pools and Fish Use of Estuarine Habitat," *Wetlands* 32, no. 3 (2012): 401–10.

22. Marisa M. Parish, "Beaver Bank Lodge Use, Distribution and Influence on Salmonid Rearing Habitats in the Smith River, California" (master's thesis, Humboldt State University, 2016).

23. Paul S. Kemp et al., "Qualitative and Quantitative Effects of Reintroduced Beavers on Stream Fish," *Fish and Fisheries* 13, no. 2 (2012): 158–81.

24. Eric Jay Dolin, *Fur, Fortune, and Empire* (New York: W. W. Norton, 2010), 286.

25. John Phillip Reid, *Contested Empire* (Norman: University of Oklahoma Press, 2002).

26. Reid, *Contested Empire*, 44.

27. Reid, *Contested Empire*, 39.

28. Reid, *Contested Empire*, 19.

29. Reid, *Contested Empire*, 34.

30. Reid, *Contested Empire*, 81.

31. Jennifer Ott, "'Ruining' the Rivers in the Snake Country: The Hudson's Bay Company's Fur Desert Policy," *Oregon Historical Quarterly* 104, no. 2 (2003): 179.

32. Ott, "'Ruining' the Rivers in the Snake Country," 182.

33. Ott, "'Ruining' the Rivers in the Snake Country," 179.

34. Reid, *Contested Empire*, 47.

35. Reid, *Contested Empire*, 175.

36. P. W. Schaffer, "Beaver on Trial," *Soil Conservation Service* (1941), as cited in Suzanne C. Fouty, "Euro-American Beaver Trapping and Its Long-Term Impact on Drainage Network Form and Function, Water Abundance, Delivery, and System Stability," in *Riparian Research and Management: Past, Present, Future* (US Forest Service, 2018), 1:102–33.

37. Tim J. Beechie, Michael M. Pollock, and S. Baker, "Channel Incision, Evolution and Potential Recovery in the Walla Walla and Tucannon River Basins, Northwestern USA," *Earth Surface Processes and Landforms* 33, no. 5 (2008): 789.

38. Beechie, Pollock, and Baker, "Channel Incision, Evolution and Potential Recovery in the Walla Walla and Tucannon River Basins," 797.

39. Daniel E. Kroes and Christopher W. Bason, "Sediment-Trapping by Beaver Ponds in Streams of the Mid-Atlantic Piedmont and Coastal Plain, USA," *Southeastern Naturalist* 14, no. 3 (2015): 577–95.

40. Rick Demmer and Robert L. Beschta, "Recent History (1988–2004) of Beaver Dams Along Bridge Creek in Central Oregon," *Northwest Science* 82, no. 4 (2008): 309–18.

41. Michael M. Pollock, Timothy J. Beechie, and Chris E. Jordan, "Geomorphic Changes Upstream of Beaver Dams in Bridge Creek, an Incised Stream Channel in the Interior Columbia River Basin, Eastern Oregon," *Earth Surface Processes and Landforms* 32, no. 8 (2007): 1174–85.

42. Nicolaas Bouwes, Nicholas Weber, Chris E. Jordan, W. Carl Saunders, Ian A. Tattam, Carol Volk, Joseph M. Wheaton, and Michael M. Pollock, "Ecosystem Experiment Reveals Benefits of Natural and Simulated Beaver Dams to a Threatened Population of Steelhead (*Oncorhynchus mykiss*)," *Scientific Reports* 6 (2016): 4.

43. Bouwes, Weber, Jordan, Saunders, Tattam, Volk, Wheaton, and Pollock, "Ecosystem Experiment Reveals Benefits of Natural and Simulated Beaver Dams," 7.

44. Nicholas Weber, Nicolaas Bouwes, Michael M. Pollock, Carol Volk, Joseph M. Wheaton, Gus Wathen, Jacob Wirtz, and Chris E. Jordan, "Alteration of Stream Temperature by Natural and Artificial Beaver Dams," *PLoS One* 12, no. 5 (2017).

45. Lewis E. Aubury, "Gold Dredging in California," *Bulletin No. 57*, California State Mining Bureau (Sacramento: 1910), 221.

46. Marina Finn, "Leave It to Beavers," *OnEarth Magazine*, February 16, 2015, https://www.nrdc.org/onearth/leave-it-beavers.

CHAPTER SIX: CALIFORNIA STREAMING

1. Heidi Perryman, "A Modest [Beaver] Proposal," *Worth a Dam*, April 16, 2017, http://www.martinezbeavers.org/wordpress/2017/04/14/a-modest-beaver-proposal.

2. "California Water 101," Water Education Foundation, 2016, http://www.watereducation.org/photo-gallery/california-water-101.

3. Joan Didion, "Holy Water," *Esquire Magazine*, December 1977, 73.

4. City council meeting minutes, City of Martinez, November 7, 2007, 1–9, http://www.cityofmartinez.org/civicax/filebank/blobdload.aspx?blobid=4828.

5. Carolyn Jones, "Martinez Mural Artist Forced to Remove Beaver," *San Francisco Chronicle*, September 30, 2011, http://www.sfgate.com/bayarea/article/Martinez-mural-artist-forced-to-remove-beaver-2298540.php.

6. Christopher W. Lanman et al., "The Historical Range of Beaver (*Castor canadensis*) in Coastal California: An Updated Review of the Evidence," *California Fish and Game* 99, no. 4 (2013): 193–221.

7. Richard B. Lanman et al., "The Historical Range of Beaver in the Sierra Nevada: A Review of the Evidence," *California Fish and Game* 98, no. 2 (2012): 65–80.

8. Michelle Nijhuis, "Evolve or Die," *The Last Word on Nothing*, February 20, 2012, http://www.lastwordonnothing.com/2012/02/20/evolve-or-die.

9. Joseph Grinnell, Joseph M. Dixon, and Jean M. Linsdale, *Fur-Bearing Mammals of California*, vol. 2 (Berkeley: University of California Press, 1937), 629–727.

10. Donald T. Tappe, *The Status of Beavers in California*, State of California Division of Fish and Game (Sacramento: California State Printing Office, 1942), 14.

11. Grinnell, Dixon, and Linsdale, *Fur-Bearing Mammals of California*, 2:635.

12. Daniel F. Williams, "Mammalian Species of Special Concern in California," State of California, the Resources Agency, Department of Fish and Game (1986).

13. Eric Jay Dolin, *Fur, Fortune, and Empire* (New York: W. W. Norton, 2010), 237.

14. Dolin, *Fur, Fortune, and Empire*, 133–65.

15. Lanman et al., "The Historical Range of Beaver (*Castor canadensis*) in Coastal California," 200.

16. Daniel Pauly, "Anecdotes and the Shifting Baselines Syndrome of Fisheries," *Trends in Ecology and Evolution* 10, no. 10 (1995): 430.

17. Travis Longcore, Catherine Rich, and Dietland Müller-Schwarze, "Management by Assertion: Beavers and Songbirds at Lake Skinner (Riverside County, California)," *Environmental Management* 39, no. 4 (2007): 460.

18. Charles D. James and Richard B. Lanman, "Novel Physical Evidence That Beaver Historically Were Native to the Sierra Nevada," *California Fish and Game* 98, no. 2 (2012): 129–32.

19. Kate Lundquist and Brock Dolman, "Beaver in California: Creating a Culture of Stewardship," Occidental Arts and Ecology Center WATER Institute, 2016, 20.

20. "Groundwater Basins Subject to Critical Conditions of Overdraft," Bulletin 118, Interim Update 2016, California Department of Water Resources, http://www.water.ca.gov/groundwater/sgm/pdfs/COD-basins_2016_Dec19.pdf.

21. Glen Martin, "A Deep Dive into California's Recurring Drought Problem," *California Magazine*, January 3, 2018, https://alumni.berkeley.edu/california-magazine/just-in/2018-01-03/deep-dive-californias-recurring-drought-problem.

22. A. Park Williams et al., "Contribution of Anthropogenic Warming to California Drought During 2012–2014," *Geophysical Research Letters* 42, no. 16 (2015): 6819–28.

23. Colin J. Whitfield et al., "Beaver-Mediated Methane Emissions: The Effects of Population Growth in Eurasia and the Americas," *Ambio* 44, no. 1 (2015): 7–15.

24. Padma Nagappan, "The Latest Climate Change Threat: Beavers," *TakePart*, January 6, 2015, http://www.takepart.com/article/2015/01/06/latest-climate-change-threat-beavers-0.

25. Ellen Wohl, "Landscape-Scale Carbon Storage Associated with Beaver Dams," *Geophysical Research Letters* 40, no. 14 (2013): 3631–36; Average carbon storage in American forests derived from Richard A. Birdsey, "Carbon Storage and Accumulation in American Forest Ecosystems," US Department of Agriculture Forest Service (August 1992): 3, https://www.nrs.fs.fed.us/pubs/gtr/gtr_wo059.pdf.

26. Executive Summary, "Sierra Meadows Strategy," *Sierra Meadows Partnership*, October 2016, https://meadows.ucdavis.edu/files/Sierra_Meadow_Strategy_4pager_shareable20161118.pdf.

27. Heidi Perryman, "Phoenix Beavers," YouTube.com, August 13, 2017, https://www.youtube.com/watch?v=3G38WTPGT4s.

CHAPTER SEVEN: MAKE THE DESERT BLOOM

1. Theodore Roosevelt, quoted in Roger DiSilvestro, *Theodore Roosevelt in the Badlands* (New York: Walker, 2011), 3.

2. Timothy Egan, *The Big Burn: Teddy Roosevelt and the Fire that Saved America* (Boston and New York: Houghton Mifflin Harcourt, 2009), 22.

3. Edmund Morris, *The Rise of Theodore Roosevelt* (New York: Random House, 2001), 387.

4. Morris, *The Rise of Theodore Roosevelt*, 388.

5. Jonathan Knutson, "Nature's Engineers? Or Nature's Despoilers?" *AgWeek*, April 11, 2017, https://www.agweek.com/opinion/columns/4248873-natures-engineers-or -natures-despoilers.

6. Gustavus Hines, *A Voyage Round the World: With a History of the Oregon Mission* (Buffalo, NY: George H. Derby, 1850), 338.

7. Tony Svejcar, "The Northern Great Basin: A Region of Continual Change," *Rangelands* 37, no. 3 (2015): 116.

8. Edward M. Hanks, quoted in Carolyn Dufurrena, "Rough and Beautiful Places," *Range Magazine*, Summer 2016, 23.

9. Warren P. Clary and Bert F. Webster, "Riparian Grazing Guidelines for the Intermountain Region," *Rangelands* 12, no. 4 (1990): 209.

10. John Zablocki and Zeb Hogan, "Partnership Protects America's Largest Native Trout in Dry Nevada," *National Geographic*, April 7, 2014, https://blog.nationalgeographic .org/2014/04/07/partnership-protects-americas-largest-native-trout-in-dry-nevada.

11. Caitlin Lilly, "Here's How Land Is Used by the Federal Government in Nevada," *Las Vegas Review-Journal*, February 18, 2016, https://www.reviewjournal.com/news/heres -how-land-is-used-by-the-federal-government-in-nevada.

12. Florence Williams, "The Shovel Rebellion," *Mother Jones*, January–February 2001, http://www.motherjones.com/politics/2001/01/shovel-rebellion.

13. Kurt A. Fesenmyer, Robin Bjork, and Teddy Langhout, "Characterizing Changes in Riparian Condition in Susie Creek Allotments: 1985–2013," Trout Unlimited Science Program. Unpublished. Accessed on ResearchGate.

14. Alan Newport, "In Defense of Beavers," *Beef Producer*, March 30, 2017, http://www .beefproducer.com/soil-health/defense-beavers.

15. Joseph Grinnell, Joseph M. Dixon, and Jean M. Linsdale, *Fur-Bearing Mammals of California*, vol. 2 (Berkeley: University of California Press, 1937), 710–11.

16. Glynnis Hood, *The Beaver Manifesto* (Victoria, BC: Rocky Mountain Books, 2011), 45.

17. Hood, *Beaver Manifesto*, 47.

18. Hood, *Beaver Manifesto*, 47.

19. Paul Greenberg, "A River Runs Through It," *American Prospect*, May 22, 2013, http:// prospect.org/article/river-runs-through-it.

20. Sandra Postel, *Replenish* (Washington, DC: Island Press, 2017), 170.

21. David L. Correll, Thomas E. Jordan, and Donald E. Weller, "Beaver Pond Biogeochemical Effects in the Maryland Coastal Plain," *Biogeochemistry* 49, no. 3 (2000): 217–39.

22. Robert J. Naiman and Jerry M. Melillo, "Nitrogen Budget of a Subarctic Stream Altered by Beaver (*Castor canadensis*)," *Oecologia* 62, no. 2 (1984): 150–55.

23. Sarah A. Muskopf, "The Effect of Beaver (*Castor canadensis*) Dam Removal on Total Phosphorus Concentration in Taylor Creek and Wetland, South Lake Tahoe, California" (PhD diss., Humboldt State University, 2007), 23–24.

24. Julia G. Lazar et al., "Beaver Ponds: Resurgent Nitrogen Sinks for Rural Watersheds in the Northeastern United States," *Journal of Environmental Quality* 44, no. 5 (2015): 1684–93.

25. Kaine Korzekwa, "Beavers Take a Chunk Out of Nitrogen in Northeast Rivers," *American Society of Agronomy*, October 21, 2015, https://www.agronomy.org/science -news/beavers-take-chunk-out-nitrogen-northeast-rivers.

26. William deBuys, *Enchantment and Exploitation*, rev. ed. (Albuquerque, NM: University of New Mexico Press, 2015), 82–83.

27. Marc Simmons, "Why Early New Mexico Just Turned from Cattle to Sheep," *Santa Fe New Mexican*, April 15, 2016, http://www.santafenewmexican.com/news/local_news /why-early-new-mexico-turned-from-cattle-to-sheep/article_a70446e1-d775-5be5 -afaa-34a3e7c811e5.html.

28. Brian A. Small, Jennifer K. Frey, and Charlotte C. Gard, "Livestock Grazing Limits Beaver Restoration in Northern New Mexico," *Restoration Ecology* 24, no. 5 (2016): 646–55.

29. Rebecca Moss, "Beavers: Nuisance or Necessary?" *Santa Fe New Mexican*, January 16, 2016, http://www.santafenewmexican.com/life/features/beavers-nuisance-or -necessary/article_50c67e90-f75d-508b-9757-81de7efeee30.html.

30. Reid Wilson, "Western States Worry Decision on Bird's Fate Could Cost Billions in Development," *Washington Post*, May 11, 2014, https://www.washingtonpost.com /blogs/govbeat/wp/2014/05/11/western-states-worry-decision-on-birds-fate-could -cost-billions-in-development.

31. Jodi Peterson, "The Endangered Species Act's Biggest Experiment," *High Country News*, August 17, 2015, http://www.hcn.org/issues/47.14/biggest-experiment-endangered -species-act-sage-grouse.

32. Darryl Fears, "Decision Not to List Sage Grouse as Endangered Is Called Life Saver by Some, Death Knell by Others," *Washington Post*, September 22, 2015, https://www .washingtonpost.com/news/energy-environment/wp/2015/09/22/fewer-than-500000 -sage-grouse-are-left-the-obama-administration-says-they-dont-merit-federal-protection.

33. J. Patrick Donnelly et al., "Public Lands and Private Waters: Scarce Mesic Resources Structure Land Tenure and Sage Grouse Distributions," *Ecosphere* 7, no. 1 (2016): 1–15.

34. Anon., "Elko County Declares Emergency After Earthen Dam Fails," Associated Press, reprinted in *Las Vegas Sun*, February 9, 2017, https://lasvegassun.com/news/2017 /feb/09/elko-county-declares-emergency-after-earthen-dam-f.

35. Ian Evans, "Oroville Dam Incident Explained: What Happened, Why and What's Next," *Water Deeply*, June 14, 2017, https://www.newsdeeply.com/water/articles/2017/06/14 /oroville-dam-incident-explained-what-happened-why-and-whats-next.

36. Marc Reisner, *Cadillac Desert: The American West and Its Disappearing Water* (New York: Penguin, 1993), 111.

37. "Dams: Infrastructure Report Card," American Society of Civil Engineers, 2017, https:// www.infrastructurereportcard.org/wp-content/uploads/2017/01/Dams-Final.pdf.

38. John McPhee, *Encounters with the Archdruid* (New York: Farrar, Straus and Giroux, 1971), 158.

39. Chris Mooney, "Reservoirs Are a Major Source of Greenhouse Gas Emissions, Scientists Say," *Washington Post*, September 28, 2016, https://www.washingtonpost.com/news /energy-environment/wp/2016/09/28/scientists-just-found-yet-another-way-that -humans-are-creating-greenhouse-gases.

40. Emily Benson, "As Sediment Builds, One Dam Faces Its Comeuppance," *High Country News*, October 20, 2017, http://www.hcn.org/articles/water-as-sediment-builds-a -colorado-dam-faces-its-comeuppance-paonia-reservoir.

41. James Pattie, *The Personal Narrative of James O. Pattie of Kentucky* (Cincinnati: John H. Wood, 1831), https://user.xmission.com/~drudy/mtman/html/pattie/pattie.html.

42. Pattie, *The Personal Narrative of James O. Pattie of Kentucky*.

43. Ken Ritter and Dan Elliott, "U.S.: 'Zero' Chance of Colorado River Water Shortage in 2018," Associated Press, August 15, 2017, https://www.usnews.com/news/business /articles/2017-08-15/us-zero-chance-of-colorado-river-water-shortage-in-2018.

44. The Lands Council, "The Beaver Solution: An Innovative Solution for Water Storage and Increased Late Summer Flows in the Columbia River Basin," completed under Washington Department of Ecology Grant #G0900156.

45. Konrad Hafen and William W. Macfarlane, "Can Beaver Dams Mitigate Water Scarcity Caused by Climate Change and Population Growth?" *StreamNotes*, USDA, November 2016, 1–5.

CHAPTER EIGHT: WOLFTOPIA

1. Michael Milsten, "From Jan. 13, 1995: Return to Yellowstone: Wolves Finally Taste Freedom," *Billings Gazette*, January 13, 2015, http://billingsgazette.com/news/state -and-regional/montana/from-jan-return-to-yellowstone-wolves-finally-taste-freedom /article_69b6adf2-57cb-57ba-b6d1-9bd9ce2a7e49.html.

2. Elliott D. Woods, "The Fight Over the Most Polarizing Animal in the West," *Outside Magazine*, January 20, 2015, https://www.outsideonline.com/1928836/fight-over -most-polarizing-animal-west.

3. Sustainable Human, "How Wolves Change Rivers," YouTube.com, February 13, 2014, https://www.youtube.com/watch?v=ysa5OBhXz-Q.

4. Arthur Middleton, "Is the Wolf a Real American Hero?" *New York Times*, March 9, 2014, https://www.nytimes.com/2014/03/10/opinion/is-the-wolf-a-real-american -hero.html.

5. Walter DeLacy, "A Trip up the South Snake River in 1863," *Contributions to the Historical Society of Montana* 1 (1876): 134, https://archive.org/stream/contributionstohvol1hist 1876rich/contributionstohvol1hist1876rich_djvu.txt.

6. Wyndham Thomas Wyndham-Quin, *The Great Divide* (London: Chatto and Windus, Piccadilly, 1876), 72.

7. Philetus Norris, *Annual Report of the Superintendent of the Yellowstone National Park to the Secretary of the Interior for the Year 1880* (Washington, DC: Government Printing Office, 1881), 43.

8. Edward Warren, "A Study of the Beaver in the Yancey Region of Yellowstone National Park," *Roosevelt Wild Life Annals* 1, nos. 1–2 (1926): 17.

9. Warren, "A Study of the Beaver in the Yancey Region of Yellowstone National Park," 23.

10. Milton P. Skinner, "The Predatory and Fur-Bearing Animals of the Yellowstone National Park," *Roosevelt Wild Life Bulletins* 4, no. 2 (1927): 205.

11. Robert James Jonas, "A Population and Ecological Study of the Beaver (*Castor canadensis*) of Yellowstone National Park" (master's thesis, University of Idaho, 1955), 165.

NOTES

12. Vernon Bailey, "Directions for the Destruction of Wolves and Coyotes," Circular No. 55, USDA Bureau of Biological Survey, April 17, 1909, 5.

13. Vernon Bailey, quoted in Alston Chase, *Playing God in Yellowstone* (New York: Harcourt Brace, 1986), 122.

14. Chase, *Playing God in Yellowstone*, 123.

15. Chase, *Playing God in Yellowstone*, 124.

16. George M. Wright and Ben H. Thompson, quoted in Chase, *Playing God in Yellowstone*, 23.

17. A. David M. Latham, M. Cecilia Latham, Kyle H. Knopff, Mark Hebblewhite, and Stan Boutin, "Wolves, White-Tailed Deer, and Beaver: Implications of Seasonal Prey Switching for Woodland Caribou Declines," *Ecography* 36, no. 12 (2013): 1276–90.

18. Thomas D. Gable et al., "Where and How Wolves Kill Beavers," *PLoS One* 11, no. 12 (2016).

19. Warren, "A Study of the Beaver in the Yancey Region of Yellowstone National Park," 183.

20. Jonas, "A Population and Ecological Study of the Beaver," 181.

21. Jordan Lofthouse, Randy T. Simmons, and Ryan M. Yonk, "Manufacturing Yellowstone: Political Management of an American Icon," Institute of Political Economy at Utah State University, August 2016, 18.

22. Douglas W. Smith and Daniel B. Tyers, "The Beavers of Yellowstone," *Yellowstone Science* 16, no. 3 (2008): 11.

23. Jennifer S. Holland, "The Wolf Effect," *National Geographic*, October 2004, http://ngm .nationalgeographic.com/ngm/0410/resources_geo.html.

24. Jim Robbns, "Hunting Habits of Wolves Change Ecological Balance in Yellowstone," *New York Times*, October 18, 2005, http://www.nytimes.com/2005/10/18/science /earth/hunting-habits-of-wolves-change-ecological-balance-in.html.

25. Rick Bass, "Wolf Palette," *Orion Magazine*, July–August 2005, https://orionmagazine .org/article/wolf-palette/

26. Warren, "A Study of the Beaver in the Yancey Region of Yellowstone National Park," 81.

27. Kristin N. Marshall, N. Thompson Hobbs, and David J. Cooper, "Stream Hydrology Limits Recovery of Riparian Ecosystems After Wolf Reintroduction," *Proceedings of the Royal Society B* 280, no. 1756 (2013).

28. Marshall, Hobbs, and Cooper, "Stream Hydrology Limits Recovery of Riparian Ecosystems After Wolf Reintroduction," 5.

29. Robert L. Beschta and William J. Ripple, "Riparian Vegetation Recovery in Yellowstone: The First Two Decades After Wolf Reintroduction," *Biological Conservation* 198 (2016): 93–103.

30. Daniel R. MacNulty et al., "The Challenge of Understanding Northern Yellowstone Elk Dynamics after Wolf Reintroduction," *Yellowstone Science* 24, no. 1 (2016): 25–33.

31. Suzanne Fouty, "Current and Historic Stream Channel Response to Changes in Cattle and Elk Grazing Pressure and Beaver Activity" (PhD diss., University of Oregon, 2003).

32. Andrew Theen, "Report: Oregon's Wolf Population Stagnant, 7 Animals Killed in 2016," *The Oregonian/Oregon Live*, April 11, 2017, http://www.oregonlive.com /environment/index.ssf/2017/04/oregons_latest_report_shows_sl.html.

33. Suzanne Fouty, "Spheres of Influence, Spheres of Impact: Preparing for Climate Change on Public Lands with New Partners, New Strategies," presentation delivered at State of the Beaver conference, Canyonville, Oregon, February 23, 2017.

34. Lina E. Polvi and Ellen Wohl, "The Beaver Meadow Complex Revisited – The Role of Beavers in Post-Glacial Floodplain Development," *Earth Surface Processes and Landforms* 37, no. 3 (2012): 335.

35. Anon., "Beavers," Rocky Mountain National Park, March 28, 2015, https://www.nps .gov/romo/learn/nature/beavers.htm.

CHAPTER NINE: ACROSS THE POND

1. Enos Mills, *In Beaver World* (Boston and New York: Houghton Mifflin, 1913), 15.

2. Róisín Campbell-Palmer et al., *The Eurasian Beaver Handbook* (Exeter, UK: Pelagic Publishing, 2015), Kindle edition.

3. Matt Pomroy, "Radioactive Wolves," *Esquire Middle East*, January 25, 2015, http:// www.esquireme.com/brief/radioactive-wolves.

4. Derek Gow, "Beavers in Britain's Past," presentation delivered at State of the Beaver conference, Canyonville, Oregon, February 23, 2017.

5. Bryony Coles, *Beavers in Britain's Past* (Oxford, UK: Oxbow Books, 2006), 72.

6. Coles, *Beavers in Britain's Past*, 73.

7. Coles, *Beavers in Britain's Past*, 159.

8. Coles, *Beavers in Britain's Past*, 139–59.

9. Mark Kurlansky, *The Basque History of the World* (New York: Walker, 1999), 48.

10. Duncan Halley, Frank Rosell, and Alexander Saveljev, "Population and Distribution of Eurasian Beaver (*Castor fiber*)," *Baltic Forestry* 18, no. 1 (2012): 171–72.

11. Duncan J. Halley and Frank Rosell, "The Beaver's Reconquest of Europe: Status, Population Development, and Management of a Conservation Success," *Mammal Review* 32, no. 3 (2002): 154.

12. Halley and Rosell, "The Beaver's Reconquest of Europe," 154.

13. Dietland Müller-Schwarze, *The Beaver: Its Life and Impact*, 2nd ed. (Ithaca, NY: Cornell University Press, 2011), 121.

14. Halley and Rosell, "The Beaver's Reconquest of Europe," 158.

15. Pearly Jacob, "Mongolia: Ulaanbaatar Signs Up Nature's Engineers to Restore River," EurasiaNet.org, August 17, 2012, http://www.eurasianet.org/node/65797.

16. Martin Gaywood, "Beavers in Scotland: A Report to the Scottish Government," Scottish Natural Heritage, June 2015, 5.

17. Anon., "Tayside Is Home to About 150 Beavers, Report Says," BBC, December 19, 2012, http://www.bbc.com/news/uk-scotland-tayside-central-20781407.

18. Tom Ough, "Beavers Are Back and Thriving But Not Everyone Is Happy," *The Telegraph*, January 22, 2017, http://www.telegraph.co.uk/news/2017/01/22/beavers -back-thriving-not-everyone-happy.

19. Jim Crumley, "Madness Over Beaver Fears," *The Courier*, November 9, 2016, https:// www.thecourier.co.uk/fp/opinion/jim-crumley/310567/madness-over-beaver-fears.

20. Katherine Sutherland, "Meet the Laird and Lady of Beaver Castle," *The Scottish Mail*, January 15, 2017, https://www.pressreader.com/uk/the-scottish-mail-on-sunday /20170115/282222305452419.

21. Anon., "Legal Challenge Over River Tay's Wild Beavers," BBC, March 2, 2011, http://www.bbc.com/news/uk-scotland-tayside-central-12612946.

22. Severin Carrell, "Scotland's Beaver Trappling Plan Has Wildlife Campaigners Up in Arms," *The Guardian*, November 25, 2010, https://www.theguardian.com /environment/2010/nov/25/beavers-scotland-conservation.

23. Anon., "Re-homed 'Tay Beaver' Dies at Edinburgh Zoo," BBC, April 1, 2011, http:// www.bbc.co.uk/news/mobile/uk-scotland-tayside-central-12934538.

24. Gaywood, "Beavers in Scotland," 18–107.

25. Gaywood, "Beavers in Scotland," 24.

26. Anon., "Beavers to Remain in Scotland," Newsroom, Scottish Government, November 24, 2016, https://news.gov.scot/news/beavers-to-remain-in-scotland.

27. Ilona Amos, "New Beavers to Be Set Free in Argyll," *The Scotsman*, October 3, 2017, https://www.scotsman.com/news/new-beavers-to-be-set-free-in-argyll-1-4576745.

28. Robert Macfarlane, *Landmarks* (London: Penguin Random House UK, 2016), 5.

29. Jessica Aldred, "Wild Beavers Seen in England for the First Time in Centuries," *The Guardian*, February 27, 2014, https://www.theguardian.com/environment/2014 /feb/27/wild-beavers-england-devon-river.

30. Anon., "River Otter Beavers 'Native to UK,' Tests Find," BBC, March 19, 2015, http://www.bbc.com/news/uk-england-devon-31971066.

31. Anon., "Research Team to Monitor Impact of Wild Beavers on Our Waterways," Phys. org, January 30, 2015, https://phys.org/news/2015-01-team-impact-wild-beavers -waterways.html.

32. Alex Riley, "Beavers Are Back in the UK and They Will Reshape the Land," BBC, October 6, 2016, http://www.bbc.com/earth/story/20161005-beavers-are-back-in -the-uk-and-they-will-reshape-the-land.

33. Mark Elliott et al., "Beavers — Nature's Water Engineers," Devon Wildlife Trust, 2017.

34. Alan Puttock et al., "Eurasian Beaver Activity Increases Water Storage, Attenuates Flow and Mitigates Diffuse Pollution from Intensively-Managed Grasslands," *Science of the Total Environment* 576 (2017): 430–43.

35. Rebecca Burn-Callender, "UK Flooding: Cost of Damage to Top £5 Bn but Many Homes and Businesses Underinsured," *The Telegraph*, December 28, 2015, http:// www.telegraph.co.uk/finance/economics/12071604/UK-flooding-cost-of-damage -to-top-5bn-but-many-homes-and-businesses-underinsured.html.

36. Jan Nyssen, Jolien Pontzeele, and Paolo Billi, "Effect of Beaver Dams on the Hydrology of Small Mountain Streams: Example from the Chevral in the Ourthe Orientale Basin, Ardennes, Belgium," *Journal of Hydrology* 402, nos. 1–2 (2011): 92–102.

37. Anon., "Beaver Wood Chip," Devon Wildlife Trust, accessed January 6, 2018, http:// devonwildlifetrust.org/shop/product/beaver-wood-chip.

38. Patrick Barkham, "UK to Bring Back Beavers in First Government Flood Reduction Scheme of Its Kind," *The Guardian*, December 12, 2017, https://www.theguardian.com /environment/2017/dec/12/uk-to-bring-back-beavers-in-first-government-flood -reduction-scheme-of-its-kind.

39. Susan Scott Parrish, "The Great Mississippi Flood of 1927 Laid Bare the Divide Between the North and the South," *Smithsonian* (first published at *Zocalo Public Square*), April 11, 2017, http://www.smithsonianmag.com/history/devastating-mississippi-river-flood -uprooted-americas-faith-progress-180962856/#WzkHszr3TAyqVUqj.99.

40. Eric Levenson, "Mississippi River Cresting in Flood-Hit Illinois, Southern Missouri," CNN, May 7, 2017, http://www.cnn.com/2017/05/06/us/flooding-mississippi-river/index.html.

41. Donald L. Hey and Nancy S. Philippi, "Flood Reduction Through Wetland Restoration: The Upper Mississippi River as a Case History," *Restoration Ecology* 3, no. 1 (1995): 14.

42. Tom Bawden, "Nazi Super Cows: British Farmer Forced to Destroy Half His Murderous Herd of Bio-engineered Heck Cows After They Try to Kill Staff," *The Independent*, January 5, 2015, http://www.independent.co.uk/environment/british-farmer-forced -to-turn-half-his-murderous-herd-of-nazi-cows-into-sausages-9958988.html.

43. Róisín Campbell-Palmer et al., "*Echinococcus multilocularis* Detection in Live Eurasian Beavers (*Castor fiber*) Using a Combination of Laparoscopy and Abdominal Ultrasound Under Field Conditions," *PLoS One* 10, no. 7 (2015).

44. George Monbiot, *Feral: Rewilding the Air, the Sea, and Human Life* (New York: Penguin Books, 2014), 5.

45. Ian Carter, Jim Foster, and Leigh Lock, "The Role of Animal Translocations in Conserving British Wildlife: An Overview of Recent Work and Prospects for the Future," *EcoHealth* 14, no. S1 (2016): 7–15.

46. Keith McLeod, "The Lynx Effect: Pilot Scheme Could See 250 of the Once Extinct Cats Back Among Scottish Wildlife," *Daily Record*, August 10, 2017, http://www.daily record.co.uk/news/scottish-news/lynx-effect-pilot-scheme-could-10959376.

47. Anon., "Beavers Are Back in Cornwall!" Cornwall Wildlife Trust, June 19, 2017, http:// www.cornwallwildlifetrust.org.uk/news/2017/06/19/beavers-are-back-cornwall.

CHAPTER TEN: LET THE RODENT DO THE WORK

1. Elijah Portugal, Joe M. Wheaton, and Nicholas Bouwes, "Spring Creek Wetland Area Adaptive Beaver Management Plan," prepared for Walmart Stores Inc. and the City of Logan, 2015.

2. US Environmental Protection Agency, Office of Water and Office of Research and Development, "National Rivers and Streams Assessment 2008–2009: A Collaborative Survey," Washington, DC, March 2016.

3. Joshua Zaffos, "'Restoration Cowboy' Goes Against the Flow," *High Country News*, November 10, 2003, http://www.hcn.org/issues/262/14362.

4. David L. Rosgen, "The Cross-Vane, W-Weir and J-Hook Vane Structures . . . Their Description, Design and Application for Stream Stabilization and River Restoration," Wildland Hydrology, http://www.creekman.com/assets/rosgen-weirs.pdf.

5. David Malakoff, "The River Doctor," *Science* 305, no. 5686 (2004): 937–39.

6. Robert C. Walter and Dorothy J. Merritts, "Natural Streams and the Legacy of Water-Powered Mills," *Science* 319, no. 5861 (2008): 299–304.

7. Timothy J. Beechie et al., "Process-Based Principles for Restoring River Ecosystems," *BioScience* 60, no. 3 (2010): 209–22.

8. Alan Kasprak et al., "The Blurred Line Between Form and Process: A Comparison of Stream Channel Classification Frameworks," *PLoS One* 11, no. 3 (2016).

9. Hal Herring, "Can We Make Sense of the Malheur Mess?" *High Country News*, February 12, 2016, http://www.hcn.org/articles/malheur-occupation-oregon-ammon-bundy -public-lands-essay.

10. William Finley, quoted in Nancy Langston, *Where Land and Water Meet* (Seattle: University of Washington Press, 2009), 101.

11. Langston, *Where Land and Water Meet*, 103.

12. Langston, *Where Land and Water Meet*, 106–07.

13. Langston, *Where Land and Water Meet*, 104.

14. Brian Maffly, "Leave It to Beaver? No Way, Says Salt Lake County," *Salt Lake Tribune*, April 8, 2017, http://archive.sltrib.com/article.php?id=5151499&itype=CMSID.

15. Ben Goldfarb, "Should We Put a Price on Nature?" *High Country News*, January 19, 2015, http://www.hcn.org/issues/47.1/should-we-put-a-price-on-nature.

16. Christopher Ketcham, "Grand Staircase-Escalante Was Set Up to Fail," *High Country News*, July 10, 2017, http://www.hcn.org/articles/monuments-how-grand-staircase -escalante-was-set-up-to-fail.

17. Jonathan Thompson, "A Reluctant Rebellion in the Utah Desert," *High Country News*, May 13, 2014, http://www.hcn.org/articles/is-san-juan-countys-phil-lyman-the-new -calvin-black.

18. Mary O'Brien, "The Good Beaver Do," *Salt Lake Tribune*, August 25, 2012, http:// archive.sltrib.com/article.php?id=54752950&itype=CMSID.

19. "Utah Beaver Management Plan," Utah Division of Wildlife Resources, developed in consultation with the Beaver Advisory Committee, Salt Lake City, Utah, January 6, 2010.

20. Mark Buckley et al., "The Economic Value of Beaver Ecosystem Services, Escalante River Basin, Utah," ECONorthwest, 2011.

21. Brandon Loomis, "Southern Utah Officials Nix Beaver Transplants," *Salt Lake Tribune*, August 27, 2012, http://archive.sltrib.com/article. php?id=54729906&itype=CMSID.

22. Rachelle Haddock, "Beaver Restoration Across Boundaries," Miistakis Institute, Calgary, AB, 2015, 25.

23. Joe M. Wheaton and William W. Macfarlane, "The Utah Beaver Restoration Assessment Tool: A Decision Support and Planning Tool — Manager Brief," Ecogeomorphology and Topographic Analysis Lab, Utah State University, prepared for Utah Division of Wildlife Resources, 2014, 4.

24. Wheaton and Macfarlane, "The Utah Beaver Restoration Assessment Tool," 5.

25. Todd BenDor et al., "Estimating the Size and Impact of the Ecological Restoration Economy," *PLoS One* 10, no. 6 (2017).

26. Glenn Albrecht, "The Age of Solastalgia," *The Conversation*, August 7, 2012, https:// theconversation.com/the-age-of-solastalgia-8337.

27. John Waldman, *Running Silver* (Guilford, CT: Lyons Press, 2011), 57.

28. Nick Paumgarten, "The Mannahatta Project," *The New Yorker*, October 1, 2007, https://www.newyorker.com/magazine/2007/10/01/the-mannahatta-project.

29. Barry Paddock, "Another Beaver Makes Bronx River Home — Doubles Total Beaver Population," *New York Daily News*, September 19, 2010, http://www.nydailynews.com /new-york/beaver-bronx-river-home-doubles-total-beaver-population-article-1.439691.

30. Emma Marris, *The Rambunctious Garden* (New York: Bloomsburg, 2011), 3.

31. Glenn Albrecht, "Soliphilia," *Psycho Terratica*, 2013, http://www.psychoterratica.com /soliphilia.html.

32. Linda Wertheimer, "Herring Make a Comeback in Bronx River," NPR, March 25, 2006, https://www.npr.org/templates/transcript/transcript.php?storyId=5301395.

33. Charles M. Crisafulli et al., "Amphibian Responses to the 1980 Eruption of Mount St. Helens," in *Ecological Responses to the 1980 Eruption of Mount St. Helens* (New York: Springer-Verlag, 2005): 183–97.

34. Kendra Pierre-Louis, "Beavers Emerge as Agents of Arctic Destruction," *New York Times*, December 20, 2017, https://www.nytimes.com/2017/12/20/climate/arctic-beavers -alaska.html.

35. J. B. MacKinnon, "Tragedy of the Common," *Pacific Standard*, October 17, 2017, https://psmag.com/magazine/tragedy-of-the-common.

36. Alius Ulevičius et al., "Morphological Alteration of Land Reclamation Canals by Beavers (*Castor fiber*) in Lithuania," *Estonian Journal of Ecology* 58, no. 2 (2009): 126–40.

Index

Note: Page numbers in *italics* refer to figures and illustrations. Page numbers preceded by *ci* refer to images in the color insert.

INDEX

INDEX

INDEX

About the Author

Terray Sylvester

B en Goldfarb is an award-winning environmental journalist who covers wildlife management and conservation biology. His work has been featured in *Science, Mother Jones, The Guardian, High Country News, VICE, Audubon Magazine, Modern Farmer, Orion, World Wildlife Magazine, Scientific American, Yale Environment 360,* and many other publications. He holds a master of environmental management degree from the Yale School of Forestry and Environmental Studies.

the politics and practice of sustainable living

CHELSEA GREEN PUBLISHING

Chelsea Green Publishing sees books as tools for effecting cultural change
and seeks to empower citizens to participate in reclaiming our global commons and
become its impassioned stewards. If you enjoyed reading *Eager*, please consider
these other great books related to nature and the environment.

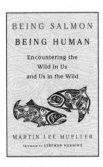

BEING SALMON, BEING HUMAN
*Encountering the Wild in Us
and Us in the Wild*
MARTIN LEE MUELLER
9781603587457
Paperback • $25.00

TAMED AND UNTAMED
Close Encounters of the Animal Kind
SY MONTGOMERY and
ELIZABETH MARSHALL THOMAS
9781603587556
Paperback • $17.95

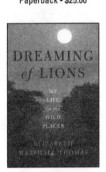

DREAMING OF LIONS
My Life in the Wild Places
ELIZABETH MARSHALL THOMAS
9781603586399
Paperback • $17.95

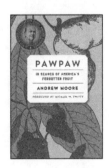

PAWPAW
In Search of America's Forgotten Fruit
ANDREW MOORE
9781603587037
Paperback • $19.95

the politics and practice of sustainable living

For more information or to request a catalog,
visit **www.chelseagreen.com** or
call toll-free **(800) 639-4099**.

the politics and practice of sustainable living

CHELSEA GREEN PUBLISHING

GODS, WASPS AND STRANGLERS
The Secret History and Redemptive Future of Fig Trees
MIKE SHANAHAN
9781603587976
Paperback • $14.95

GRASS, SOIL, HOPE
A Journey Through Carbon Country
COURTNEY WHITE
9781603585453
Paperback • $19.95

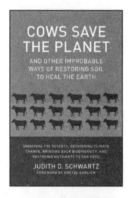

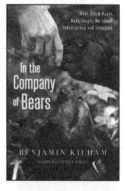

COWS SAVE THE PLANET
*And Other Improbable Ways of
Restoring Soil to Heal the Earth*
JUDITH D. SCHWARTZ
9781603584326
Paperback • $17.95

IN THE COMPANY OF BEARS
*What Black Bears Have Taught Me
about Intelligence and Intuition*
BENJAMIN KILHAM
9781603585873
Paperback • $24.95

the politics and practice of sustainable living

For more information or to request a catalog,
visit **www.chelseagreen.com** or
call toll-free **(800) 639-4099**.